RESUMES FOR THE OVER-50 JOB HUNTER

Samuel N. Ray

John Wiley and Sons, Inc.

New York ▪ Chichester ▪ Brisbane ▪ Toronto ▪ Singapore

In recognition of the importance of preserving what has been written, it is a policy of John Wiley & Sons, Inc., to have books of enduring value published in the United States printed on acid-free paper, and we exert our best efforts to that end.

This publication is designed to provide accurate and authoritative information in regard to the subject matter covered. It is sold with the understanding that the publisher is not engaged in rendering legal, accounting, or other professional services. If legal advice or other expert assistance is required, the services of a competent professional person should be sought. *From a Declaration of Principles jointly adopted by a Committee of the American Bar Association and a Committee of Publishers.*

Library of Congress Cataloging-in-Publication Data

Ray, Samuel N., 1926-
Resumes for the over-50 job hunter / by Samuel N. Ray.
p. cm.
ISBN 0-471-57422-8 (cloth)—ISBN 0-471-57423-6 (pbk.)
1. Résumés (Employment) 2. Aged—Employment—United States. 3. Middle aged persons—Employment—United States. 4. Age and employment—United States. I. Title. II. Title: Resumes for the over-fifty job hunter.
HF5383.R39 1993
808'.06665—dc20 92-22988

Printed in the United States of America

10 9 8 7 6 5 4

Acknowledgments

MY THANKS TO MY COLLABORATORS:

Gail Ryder. Gail, my Transition Team colleague, provided expert advice regarding the content of this book . . . especially in the selection of the sample resumes and letters, most of which came from her files.

Carol Turkington. Carol took my notes, outlines, ideas, and research materials . . . and worked with me to produce a book that will, I trust, provide the over-50 job seeker with immediate, practical guidance.

Preface

This book is for those of you approaching 50—and those of you who have left 50 far behind. Maybe you're out of work, or you want to reenter the job market, or you're just looking to make a change.

Whatever your reasons for being in the job market, the first thing you need to do is prepare a knockout resume. This "paper profile" gives an employer that crucial first glimpse of who you are and what you can do before you ever walk in the door. In fact, your resume can determine whether or not you get in the door at all.

Good resumes are powerful advertising tools. Bad ones—and I've seen plenty—eliminate a prospect quickly. The key to good resume writing is persuasive copy that says just enough to snag an interview. You may have already lived a long and productive life, but your resume should not be a complete history of your life and work: It's a preview of the coming attraction, not a full-length film. Ideally, you should carefully design your resume so a prospective employer wants to meet you.

There are lots of resume-writing books out there on your bookstore's shelves, so why do you need one aimed at the over-50 job hunters? Because you need a resume designed to emphasize your experience, maturity, and skills, while sidestepping the obvious age giveaways: information that will relegate your resume to the "reject" pile because you seem too old.

And make no mistake—this occurs all too often.

To make sure it doesn't happen to you, follow the suggestions in this book and pay careful attention to the sample resumes that I have provided. Keep in mind that the purpose of a resume is to get you an *interview.* A good resume will open doors to job opportunities, but it won't get you a job. Only *you* can get you a job.

Resumes for the Over-50 Job Hunter is based on my experiences in business during the past 35 years. As a human resources executive, I have read hundreds of thousands of resumes, interviewed countless applicants, and hired hundreds of candidates—from clerks to vice presidents.

I also know what it's like to be over 50 and looking for a job, since I changed jobs myself twice after my 50th birthday. Today, in my 60s, I'm president of The Transition Team, an outplacement firm headquartered in metropolitan Detroit, which helps displaced employees coast to coast find jobs. I've helped out-of-work food processors, coal miners, autoworkers, machinists, salespeople, secretaries, engineers, art directors, corporate vice presidents, and attorneys prepare resumes. Many of these people were over 50, and all were successful in finding new jobs.

In 1991, I shared my experiences in the book *Job Hunting After 50: Strategies for Success.* Now I'd like to help you concentrate on one aspect of the job search: how to prepare a resume to take advantage of your special strengths and experience.

SAMUEL N. RAY

Troy, Michigan

Contents

PART I

BEFORE YOU BEGIN . . .

1

The Preliminaries

Bob N., 52, had been an engineer with a prominent Detroit firm for 25 years when he lost his job in the wake of a corporate takeover. Not ready to retire, he felt sure he could translate his extensive experience into an exciting new position—as long as he could get his foot in the door without giving away his age. With a revamped resume designed to emphasize talents and sidestep age, Bob landed an interview and so impressed the company's president that he was hired as chief engineer within two months.

Whether a person is just out of college or a seasoned pro with 30 years of business experience, all job hunters must follow the same rules for good resumes. But there are special rules for 50-plus candidates that we must know so we don't get written off because of age.

Age? Isn't age discrimination illegal?

Yes, it is. But let's face it—age bias by employers *does* exist. Given the choice between equally qualified candidates, an employer will often vote for youth over maturity. Some employers believe that older workers are less productive than younger workers, that they are absent more often, that they are rigid and inflexible.

You and I know better. And what we know was confirmed by a 1991 study conducted by ICF Inc., a Washington, D.C., consulting firm specializing in labor research. In an overview of three large companies, ICF found that older workers have lower turnover and absenteeism rates than younger workers, and are often better salespeople. Furthermore, the study found that older workers master new technologies as quickly as younger employees and are just as flexible in accepting work assignments.

For example, at Days Inns of America, Inc., researchers discovered that it takes the same amount of time to train both younger and older workers on sophisticated computer equipment but older workers stay longer, thereby cutting overall training and recruiting costs.

The same study also found that when executives at a British retail chain in 1988 tried staffing one store entirely with workers over 50, that store turned out to be 18 percent more profitable than five comparable stores. Employee turnover at that store was six times lower, there was a lower absenteeism rate, and the number of thefts decreased.

In fact, a range of recent studies have found that:

- Learning ability, intelligence, memory, and motivation do not decline with age.
- Accident and attendance records are better for older workers than for younger employees.
- Productivity levels remain stable or increase with age for most jobs.
- Older workers experience less stress on the job and have a lower rate of admissions to psychiatric facilities than younger workers.
- Maturity is an aid to evaluating new information and making reliable, consistent decisions.
- Adaptability is unrelated to age.
- Employees over age 65 take fewer days off for illness than younger workers.

So what does this have to do with resumes?

Plenty. You need to understand that while ageism is based on myths, not facts, it still exists. And if your resume tips off your prospective employer that you are over 50, you might not get a chance to discuss your skills, your experience, your maturity, and your vitality in a face-to-face meeting. The resume is your ticket to an interview; without the interview, you can't become a viable job candidate.

THE JOB MARKET

But what about the job market? So many people tell me, "There's no jobs out there for us old timers."

That's just not true.

In the first place, America isn't a nation of youngsters anymore. Today, 60 million Americans are over age 50. That's one-fourth of the country's population—the portion that controls 55 percent of the country's income and 70 percent of its assets. And by the year 2050, the over-55 population will more than double.

In fact, it is quite possible that, someday soon, being over 50 will become fashionable, since a society tends to take on the character of its most prevalent citizens—and America's most prevalent citizens today are the baby boomers born during this country's last great population explosion.

As the 80 million 'boomers gallop into the 21st century, the whole pattern and fabric of American life is expected to change. It will change, experts predict, because numbers spell power, and no group before or since has wielded so much. The whole country watched as this population group moved from adolescent angst to the Vietnam war experience, from acquisitive early adulthood to late-30s parenthood.

In middle and old age, the continuing influence of the baby boomers in American culture is expected to have a profound effect on everything from taxes to social services, Social Security to nursing homes. One expert suggests that if the federal government continues to support people over 65 at current levels, senior citizens will be receiving 60 percent of the national budget by 2030 when the mass of baby boomers will have moved past age 65 and on into old age.

There are signs that Congress has been aware of the ramifications of the aging of the country for some time. In 1987, Congress struck down laws allowing mandatory retirement at age 70, and the Social Security Administration is piling up reserves in anticipation of the upcoming financial drain. But experts are concerned that it just won't be enough.

If we're to make a success of this greying of America, people will just have to start viewing senior citizens not as hopeless and burdensome but as useful and productive. There will have to be a place for older Americans to make their own way in the world, because the rest of the population simply won't be able to support them all.

Companies have already begun to realize the value of older workers as a solution to short-term economic problems. Just as a company may hire a temporary typist when things get busy, firms today are discovering the advantages of hiring older, experienced "consultants" for professional positions and "temps" for other work. This approach saves money by avoiding full-time salaries and benefits while still attracting top-notch workers.

Still, this doesn't mean that workers over age 50 are being actively recruited. At least through the end of this decade, early retirement is likely to be emphasized in the wake of industrial reorganizations and mergers and a lackluster national economy. It will be up to over-50 workers to go out and market themselves in such a way that companies will begin to see the wisdom of hiring older, experienced, and mature employees.

SO WHO NEEDS A RESUME?

Now that you understand the special concerns of those over 50 in today's job market today, you can see why your resume should reflect those concerns.

And make no mistake—*every* job seeker needs a resume. In the past, things were different; the only people who really needed a resume were professionals and managers. But my experience as a personnel manager and outplacement consultant has taught me the value of a resume for all job seekers, regardless of the job title and salary level.

What does a resume do for the typical hourly employee?

- ➤ It makes his or her application stand out from the crowd. I recommend that the job seeker who walks into a business to fill in an application take a little pocket stapler along. Staple a copy of the resume to the completed application.
- ➤ It's useful as a networking tool, especially with friends and relatives:

  JOB HUNTER: As you know, I'm looking for a new job. Would you speak to your employer for me?

  FRIEND: Sure.

  J.H.: Here's a copy of my resume. It will tell your employer my name, address, and phone number, as well as information about my experience and education.

So, how do you go about it?

First of all, it's important to understand what the resume is and what it isn't. The resume is not a month-by-month life and job history. It's not the story of your Korean War exploits, your hobbies, or your grandchildren. And it won't land a job for you.

The resume is a tool for getting job interviews. Every element of the resume must be examined with these questions in mind:

- ➤ Will this help me get the interview?
- ➤ What will make the employer want to meet me?
- ➤ What will be a turnoff?

You must read your resume through your prospective employer's eyes. He or she will probably be sitting at a desk piled high with hundreds of resumes, looking for a way to winnow the stack. You want to make sure your resume is saved with the grain, not thrown out with the chaff; it is a sure bet that anything sloppy or careless will end up on the threshing room floor. You can just about guarantee that resumes written by hand, printed on bright pink stationery, or typeset with a photo embossed on the upper right corner will be immediate losers.

Your aim is to avoid doing anything that would turn off an employer, and yet stand out from the pack enough to spark some interest.

WHO SHOULD WRITE IT?

Lots of people out there could do the job—for a fee. These include:

- Professional resume writer
- Computerized "resume-writing" program
- Outplacement firm
- Employment agency

If you look in the Sunday papers, you will find many ads for professional resume writing services. Some are even "guaranteed" to get you interviews. (Be forewarned: Anytime somebody guarantees to get you an interview or a job, keep your wallet in your pocket. There are no genuine guarantees in this business.)

Then there are computer programs, which are fine at cranking out a resume; the trouble is, it *looks* as if it has been cranked out of a computer. Too many professional resume writers and computer programs use the "cookie cutter" approach. All their resumes look the same—and employers recognize them. And while outplacement firms and employment agencies can help, the key word here is "help." To be most effective, you need to write your resume yourself.

To write your own resume, you will have to review your unique experience, accomplishments, and educational background. Taking this personal inventory prepares you to talk intelligently about yourself when you're out marketing your own services (that is, looking for a job).

So go ahead and get all the help and advice you want: Talk to your outplacement counselor, an employment specialist, a trusted friend. These people can help you with format, wording, or typing. But be sure the resume is your own, not a routine, professionally prepared clone like a hundred others.

So! Are you ready to start living the adventure of the second half of your life at a time when our country might really begin to respect your talents and experience? Then it is time to start crafting a resume that will convince an employer to interview *you.* I'll take you step by step through a process I've designed to help you inventory your strengths and assemble this information in a well-written resume that will convince an employer you've got what it takes.

2

The Age Issue

Robert T. had been an electrical engineer for an international home products company for 23 years when his company had a massive cutback. When he was asked about his previous job, he commented: "In the old days, you stayed with one company. Nowadays, these young hotshots are hopping from one job to another. And these young gals just out of school are given first crack at the jobs."

It is easy to see that Robert's approach was all wrong. With this outdated image of himself and his habit of talking as if the best were behind him, Robert didn't have much luck in finding anything new. It wasn't until he realized he needed to carve a few years off his attitude that he finally found a new job.

Obviously, you can't change the numbers on your birth certificate or your driver's license. But that doesn't mean you can't present a younger *attitude.* You'd be surprised how minor details can reveal a person's age and have a negative effect on an employer who's not interested in tired, old ideas. You certainly don't want to lie to anyone, but you *can* spruce up your actions, attitudes, and appearance to make a vigorous, vital impression.

REFRAME YOUR IMAGE

Psychologists call it "reframing": refocusing attention on one aspect of yourself to present a different image. When an employer asks you about yourself, don't give clues to your age. Sure, you're proud of your grandchildren; I'm proud of mine, too. And once you get the job, you can put photos of them on your desk. Just don't mention them during the interview. Also avoid references to the year you graduated, your Korean War experience, or your 1949 Chevy.

You know that you're in the prime of your life, and it's up to you to convince an employer. If you're asked about interests and activities, talk about the marathon you just won or the recent skiing vacation, about your volunteer work and civic responsibilities. You want to emphasize positive, energetic contributions. Then try your best to steer the conversation to work-related achievements.

ELIMINATE SEXISM

Robert didn't realize it, but referring to women in his profession as "young gals" delivered a message to listeners that Robert was old-fashioned and, some people would say, just plain sexist. His attitude suggested that he was questioning the ability of a woman to do her job in a "man's world." Employers tend to steer clear of anyone whose behavior suggests it could cause trouble by provoking harassment or discrimination complaints from coworkers. And since it was quite possible that Robert would be interviewed by a woman, he really needed to curtail these outdated attitudes.

However well-intentioned, never refer to a woman over age 18 as "girl," "gal," "dear," "sweetie," or "honey." And it goes without saying that you should never make any other form of prejudicial remark or negative comment on race, religion, or nationality.

CUT OUT DATED PHRASES

Robert's conversation was heavily laced with words that carry the indelible stamp of age—not to mention pessimism bordering on bitterness. The aim here is to present a positive appearance without any telltale remarks that give away your age. Robert needed to eliminate comments such as "in the old days," "nowadays," and "young hotshots."

Here are some phrases which you should try to eliminate from your vocabulary, especially during an interview:

Years ago . . .	When I was younger
When I was starting out	The gals (or girls) in the office
When I was your age	Back when . . .
When you get to be my age	In the good old days

MUM'S THE WORD WHEN IT COMES TO AGE

Because most employers know it's illegal to come right out and ask your age, most won't attempt it, although a few will take the risk, laughing self-consciously and saying "I'm not supposed to ask you this, but . . ." Others will try to figure out your age by asking leading questions that may cause you to give away your age. Being forewarned, you know better than to fall into that trap—and you know your age doesn't matter as long as you're qualified for the job.

PROJECT SELF-CONFIDENCE

The important thing is to fix the knowledge that you can do the job firmly in your mind. Don't get rattled about being "older." What does "older" mean, anyway? Just that you've had more time to gather more experience, maturity, and know-how. Therefore, you want to showcase these qualities rather than enter an interview apologizing for your age.

Never say:

- I guess you'll probably be looking for someone younger . . .
- Well, I'm 59, but I don't look it, do I?
- Who's going to hire somebody my age?

Apologizing for being over 50 plays into prejudices that unfortunately may already exist. But as long as you're qualified, age just shouldn't have anything to do with whether or not you get the job.

STAY SHARP

In today's competitive job market, you have to be physically and mentally alert to make it. The "old days" practice of three- or four-martini lunches is history. You must be aware of the changing economy and how business has changed over the past 20 years.

KEEP TECHNOLOGICALLY CURRENT

Learning is a lifelong process. Just because you left school 30 years ago doesn't mean you can ignore what's happening in your field.

No matter what your profession, you should realize that many employers believe the over-50 worker hasn't kept up with technological advances. Make sure you're up to par. It's true that the "old way" of doing things might still work, but such methods are not *always* the best just because that is the way you're used to doing things. And no matter how well you think those methods work for you, an interviewer who perceives that other candidates are more up to date, won't give you the job. It is never too late to learn what you need to know to succeed.

COMPUTER SAVVY CAN HELP

It will be invaluable if you've taken the initiative to learn at least a little bit about computers and how they work, even if you don't use a terminal on your everyday job. This is the computer age, and these machines really do improve productivity. If you've taken the time to learn how to develop and produce your own reports, spreadsheets, memos, and proposals—or at least understand how they are done—you've proven you can adapt to the changing workplace.

If your current employer offers computer training, take every course available. If not, and you don't like to ask co-workers or friends for help (and you've found trying to train yourself an impossibility), enroll in a computer class. There are plenty available in most communities, at local senior centers, adult education centers, and even computer stores. You could buy an inexpensive machine or consider renting or leasing one. Many public libraries also offer free use of computers. If you can't type, that may be your biggest obstacle to computer literacy. Fortunately, touch typing is not difficult to learn; check your local adult education center for evening courses. There are also computer programs that can teach you touch typing.

FIT AND TRIM

You want to present the best image you can to prospective employers, and being grossly out of shape and terribly unfit is not the best way to do that. Employers want workers who will be able to do the job. If you're already worried about being considered "too old," it's in your best interest to counteract that prejudice with a healthful image.

Begin with a visit to your doctor for a complete physical and ask about vitamins, diet, and exercise. Then, *follow* that advice.

Before long, you'll find you have more vitality and energy than ever. Ironically, it is often the over-50 worker—who can least afford it—who has abandoned a healthy lifestyle while pursuing a career.

And while you're working on getting back into shape, you can compensate for a less-than-perfect physique with the right kind of clothes and grooming.

YOUR CLOTHES: CURRENT BUT CONSERVATIVE

It might seem unfair, but an interviewer is already sizing you up within 30 seconds after you enter the room; and a negative impression is almost impossible to erase, no matter how sharp your verbal presentation or impressive your resume. If your clothes fit poorly, if they're outdated or garish, you'll be on your way out without even knowing it—and there's no way you can prove you were discriminated against for wearing the wrong clothes.

That is why the best idea is to dress conservatively. Check out the clothes worn by people in several positions higher than the job you are seeking, and dress accordingly. Here's what you will need for an impeccable interview wardrobe:

- Two suits or ensembles (conservative, well-tailored)
- White shirts/blouses
- Good-looking ties and accessories

When it comes to suit colors, it's best to go with solid navy or gray: Psychologists say those colors best convey an image of power and authority. The earth tones, such as brown and tan, do just the opposite and should be avoided. Women may wear suits, but a dress with a jacket is also acceptable, as long as it is well-cut, businesslike, and conservative. Sports coats and blazers are not acceptable interview clothing for men; pantsuits or flashy, suggestive clothes are a poor choice for women.

As long as we're talking about clothes, don't forget your feet. Although you may not spend a lot of time looking at your toes, your footwear is an integral part of your professional "look." For men, best bets are lace-up leather shoes such as black leather, highly polished wing tips. If you can't afford a new pair, have yours resoled and polished. You should also have a black leather belt and watchband. Women should wear dark leather pumps with no more than a two-inch heel.

Accessories

When it comes to accessories for men, eliminate flashy pinky rings or diamond pins. Men should wear only a wedding band and watch (and, perhaps, a school ring). Women should never wear any flashy, cheap jewelry or makeup. If you need help in the makeup department, visit a skin salon or a cosmetic counter at a good department store and explain you want a polished, professional look.

A woman's purse should be as understated as her shoes: good quality leather, in good condition, in the same color as shoes (and belt if worn). Many women find they present a sleeker, more "pulled-together" image if they keep bare necessities (keys, compact, wallet, lipstick, and comb) in a compartment in their attaché cases. Or you can keep a small, thin clutch purse containing those necessities inside your attaché case.

The attaché case itself is an essential element for any job seeker. You can keep a few extra copies of your resume there, samples of any work you've done that might be a plus at the interview, paper for taking notes, and pens. Make sure your attaché is as clean and presentable as the rest of your appearance.

Shop for Success

Unless you have at least two perfectly tailored outfits that are no more than three years old hanging in your closet, it's time to head for the clothing store. While quality is important, you needn't go overboard; it's possible to find brand-name clothes at factory outlets or department store sales.

Department and speciality stores sometimes offer the services of an "image consultant." By all means, utilize that professional advice. A consultant can tip you off to your most flattering styles, colors, and accessories; just make sure you stick to classic designs and avoid fads. Telling the consultant that you are interviewing for an important job should produce the right kind of advice.

When you head off to the stores, remember that you need to present a "total look." If you only need to buy a few accessories, take along your suit so that you can try on the entire outfit at the store. If you only need the suit, be sure to bring along the blouse or shirt, the tie, belt, shoes. This is a big investment, and looking your best during a job search is of critical importance.

Once you have the entire outfit together, inspect yourself in the store mirror and ask yourself:

- Does it fit well?
- Is this fit flattering?
- Is anything uncomfortable?
- Are there any shiny spots?
- How are the cuffs and hems?
- Is there enough material under the arms?
- Does the material bunch or crumple? (Try sitting and standing.)
- How do you feel wearing the outfit? Are you comfortable and confident?

Exception to the Rules

These clothing basics apply to every job *except* those in the high fashion, entertainment, or certain arts professions where people are expected to express themselves creatively through their clothes. This still doesn't mean you should show up for an interview in a shocking or daring outfit; it just allows for a bit more leeway (say, a colorful tie or scarf). It is always better to err on the conservative side no matter what job you're after.

Now that you've considered what you'll be wearing on your interview, how about your hair? The last thing you want is for it to look limp or dirty. While a company can't expect perfection, it can aim for a neat, clean, and attractive appearance—and so should you.

Men who are over 50 may be losing their hair, but it is a mistake to attempt to disguise that fact with poor-quality toupees, wigs, or unflattering dyes. If you feel more comfortable with a wig or toupee, then make sure you buy a high-quality hairpiece and have it professionally fitted.

It may seem like an awful lot of rules, but a tough, *youthful* world is waiting out there. If you want to compete with old *and* young workers, you're going to have to pay attention to those rules. Whether you like it or not, you're being evaluated—and possibly hired—by people who not only know the rules but live by them. They won't appreciate a maverick who prefers nonconformity.

3

Defeating Age Discrimination

We all know age discrimination is illegal. And we all know that it still exists in the American workplace. The good news is that workers are learning more and more about their rights and how to make sure they aren't denied a job just because they may be getting older.

AGE DISCRIMINATION IN EMPLOYMENT ACT

In an attempt to protect older Americans in the workplace from discrimination, the federal government in 1967 passed the Age Discrimination in Employment Act (ADEA), which prohibits age discrimination in workers over age 40. It is administered through the Equal Employment Opportunity Commission (EEOC), which has offices across the country. Basically, the antidiscrimination law:

- Promotes employment of older persons based on their ability rather than their age
- Helps employers and workers find ways to solve problems caused by the impact of age on employment
- Forbids arbitrary age discrimination in employment

Specifically, the law states that companies with more than 20 employees may not:

- Indicate an age preference in notices or job advertisements
- Discriminate against any worker over age 40 in hiring, firing, or any other aspect of employment because of age
- Take action against an employee for bringing a complaint of age discrimination against an employer or by helping the government investigate an alleged case of age discrimination

The law also states that employment agencies may not discriminate in referring workers over age 40.

EXCEPTIONS

There are exceptions to this law, which employers may use to defend themselves in court against charges of age discrimination. Employers may:

- Consider age as a basis for hiring when it is necessary to the normal operation of a business (for example, hiring a young actor to play a child's part in a play)
- Fire or discipline an employee for good cause
- Observe the terms of a bona fide seniority system or any bona fide employee benefit plan
- Take an action based on reasonable factors other than age
- Retire certain executives or high-ranking policy-making employees at age 65
- Retire tenured faculty at institutions of higher learning at age 70 (effective January 1, 1987, through December 31, 1993)
- Fire or refuse to hire firefighters or law enforcement officers under applicable state and local laws effective January 1, 1987, through December 31, 1993)

Other than these exceptions, the age discrimination law forbids almost any employment action involving age, even when there are other legitimate reasons for that decision. It also prohibits discrimination *within* the protected group (for example, hiring a 45-year-old over a 50-year-old on the basis of age).

In addition to the federal law, most states and many local governments also have enacted age discrimination laws. If any of these local laws provide more protection that the ADEA, the local laws take precedence.

Despite more than 20 years of legal protection against age discrimination, it still exists. But by recognizing the problem, it's possible to take steps to overcome it. The problem with age discrimination is that it is often subtle. Few employers would be naive enough to send a rejection letter telling an applicant, "You're too old." Instead, a polite "No, thanks" rejection letter does the job, and much more safely, too.

> Angela was a medical editor who sent a resume describing her qualifications to a large pharmaceutical firm in response to a classified ad. The company responded enthusiastically and expressed an interest in her background. The personnel manager enclosed an application form "just as a formality" before scheduling an interview. Instead, as soon as Angela had returned the filled-out form—loaded with requests for dates—the company sent Angela a form letter stating there were no openings suited to her qualifications.

Why the sudden change of heart? It's obvious that Angela's qualifications didn't change in one week. It's likely that the company's personnel manager took a good look at Angela's detailed work history and figured out that she was just "too old" for the job being advertised.

YOU CAN OVERCOME DISCRIMINATION

Discrimination on the basis of age can be hard to prove. Since you didn't yet have the job, it's hard to prove damages from losing it. Besides, you don't really want to spend time and money pursuing an errant company when all you're trying to do is progress in your career.

So what should Angela have done in this case? I'd have advised her to call up the personnel manager, in a very polite tone of voice, and say this: "Since my professional qualifications didn't change in between your first and second letter to me, I can only assume that some personal information on my application caused you to disqualify me from consideration. I'm still very much interested in the position, and I think I'd be an asset to your company. I'd appreciate an opportunity to present my qualifications more fully in an interview."

Sometimes, of course, employers may come right out and tell you that they think you're too old. They'll tell you that the position is an "entry level" position or designed for "recent college graduates." In that case, smile pleasantly at your interviewer and say: "I believe that the practice of hiring only workers of a certain age for a job is illegal. I would like you to consider that my experience and discretion would be an important asset to your company in this particular job."

In either case, don't expect employers to welcome you with open arms. Attitudes change slowly in the workplace, especially when it comes to age discrimination. But at least, having spoken your mind, you can feel confident that you've done all you could reasonably be expected to do. Discriminatory practices are often so ingrained in company policy that employers don't even think about whether they are being discriminatory. Pointing it out to them won't win you any points and probably won't get you the job, either. In fact, I know of *no* instance where an applicant alleged discrimination and got the job.

AVOIDING DISCRIMINATION

Instead, the best way to avoid discrimination is to get an interview without revealing your age. You can do that with the right resume, and this book will help you construct a resume that highlights your experience and skills, not the date on your birth certificate.

During an interview, if you detect serious age discrimination and you're worried about staying in the running because of it, it may be a good idea to reassure the employer. Try this:

> I have a feeling that you may be a bit concerned about my age. You know, I believe my age is really an asset in this position, not a liability. It's given me seasoning and judgment that this job calls for—and I've continued to grow during my professional life. Even if I *could* turn back the clock 10 years, I wouldn't, because those years have taught me valuable skills. I'm sure you've heard about that new research showing older workers (executives, managers, etc.) are more loyal and hardworking than younger workers, and no less productive, haven't you?

Ending this speech with a friendly question takes the sting out of the message and give the interviewer the opportunity to agree with you.

Of course, we've all met interviewers who know the law but plunge right ahead anyway, asking bluntly: "I shouldn't ask this question, but *how old are you?*"

You want to deflect this direct question without appearing rude. It's not a good idea to get huffy and scream: "You can't ask me that!" Obviously, the interviewer can—he or she just has. Objecting to the question will only be irritating and give the impression you can't handle yourself very well in a crisis. Instead, try this in response to a direct "How old are you?":

> . . . I think what you're asking me is: . . . How long will I be in this position? . . . Then pause, and say firmly: "I'm going to be here at least five years. How many younger candidates can promise you that?"

By remaining unflappable and responding in a positive, firm manner, you've evaded the direct question while showcasing your keen negotiating skills. Moreover, you're getting to the real heart of the question—whether or not you're going to be an effective player on the company team.

There are a range of other not-so-subtle questions that may pop up during an interview:

- How do you feel about working for a younger boss?
- How old are your children?
- Do you have grandchildren?

Keep in mind that these questions are not illegal per se, and it is best not to get riled when they crop up. If you filed a complaint, you'd have to prove that the question was asked for the purpose of discriminating against you.

Instead, prepare yourself for them. If you are genuinely offended by these personal questions and you decide that you don't want to work for anyone who would ask them, then refuse to answer and end the interview.

If you need the job, there are several approaches:

- Try humor: "I'm 50—twice as good as when I was 25."
- Be sincere: "I'm 55, in excellent health, and—as my resume shows—I have plenty of experience and achievements to offer."
- Reframe the question: "What I'd really like to discuss is my success in providing new customers for the Acme company."

ON-THE-JOB DISCRIMINATION

Discrimination during the hiring process is only one aspect of ageism. Once you get the job, you still may find yourself dodging ageist booby traps in the workplace. Therefore, I've also included here a discussion of discrimination on the job and how to handle it.

DISCRIMINATION IN LAYOFFS

According to the law, layoffs must be based on merit or seniority: the last-hired, first-fired approach. But what sometimes happens is that when a company is looking at the bottom line, they see those with the highest salaries are those with the most seniority. With this mind-set, it seems to make sense to fire those who've been around the longest and are earning big paychecks, and replace them with recent college graduates or even part-time help, at much lower rates of pay.

This is, however, illegal. A company that lays off a productive, older employee because of age and salary level considerations is violating federal law and can be forced to reinstate the employee with full back pay. Employers have often attempted this ploy, resulting in the payment of millions of dollars in liability. Some of the better-known suits brought by the federal government resulting in large awards for workers include:

- A research facility that laid off hundreds of engineers while recruiting at colleges for replacements
- A life insurance company that laid off 150 older sales representatives during a cutback
- A railroad that laid off every worker eligible for a pension instead of younger workers

DISCRIMINATION IN FIRING

A person can be fired for many reasons. But it is considered age discrimination if older workers are held to certain standards that are not applied to younger workers with the same characteristics.

HARASSMENT

Age discrimination can take many forms, and hassling an older worker until he or she quits so the company can hire younger workers is illegal. What constitutes "harassment" or pressure?

> Paul S. had passed his company's usual retirement age, but he was still on the job. Younger workers were constantly asking him why he didn't quit. Others would mutter comments about "the old man"

within earshot, wondering why he didn't leave and give a younger person a shot at the job.

While not practiced directly by his employer, what happened to Paul is still considered harassment. It is an employer's responsibility to make sure this doesn't happen in the workplace. If it does, the employer can be held liable. More and more older workers are taking companies to court to stop this type of harassment.

EMPLOYEE BENEFITS

While older workers may legally be offered lower benefits than younger workers in some employee benefit plans, the law doesn't grant a company the power to shut out older workers entirely. It is legal if the same life insurance premium payment provides lower benefits for older workers because of the higher risk in that group. But even workers over age 65 are entitled to the same pension and medical benefits offered to younger workers.

HOW TO FILE A COMPLAINT

To win an age discrimination suit, you must prove that you were adversely affected by an employment action and that the facts indicate age was a consideration in that action. If you think you have a case, here's what to do:

1. File a complaint with your employer (and your union, if applicable). Check with the company's human resources department to make sure you comply with any established procedures.

This may be all it takes. In the wake of a written complaint, your employer may reverse the decision. If not, you must still explore every avenue to solve the problem before you can qualify for protection by the ADEA.

2. If the company has not stopped its discrimination, check with the nearest Equal Employment Opportunity Commission office and your state government agency responsible for handling age discrimination complaints. If the state law seems more favorable to you than the federal law, or if the state agency plans to prosecute faster or more aggressively, file your complaint with the state agency before you file a

second one with the EEOC. In general, the agency who receives the complaint first will handle the case. It is best to meet personally with an official in each office.

3. You must file a second complaint with the other agency in order to protect your right to sue in court later on. If you file with the state first, you must also file a charge with the local office of the EEOC. Usually, such a charge must be filed within 180 days of the date of the alleged act of discrimination, counting from the date you received notice from the employer of an action (such as a layoff). If the discrimination has taken place over a period of time and there is no specific date, file your charge as soon as you have gathered the information that convinces you that you've been discriminated against. Your charge must be a written statement of facts in a letter addressed to the nearest EEOC office. In the letter, include your name and address, and also your employer's; your job title; a brief job description; dates of employment; and a description of the discrimination.
4. You'll receive a letter telling you to schedule an interview with an investigator who will ask you questions, look at any evidence, and discuss the remedy. Take to that meeting copies of any written documents (such as company policy manuals, handbooks, letters, memos, etc.) that help prove your case.
5. The agency will investigate and attempt to negotiate a settlement. If no settlement is reached, the agency may file a lawsuit, although this rarely happens. It is more likely that you will file your own lawsuit through a private attorney or a public assistance agency.
6. If you do file your own lawsuit, choose a lawyer who is experienced in age discrimination cases, is interested in your case, and believes that you have a chance. Make sure you find out how much it will cost and how long it will take. If you pay by the hour, set a limit on how much you're willing to pay and what you expect.

The EEOC has been criticized in the past for being understaffed, overburdened, and mismanaged. Their track record certainly is far from perfect. If you are thinking about filing a lawsuit through your private attorney, make sure you realize how much time and money it will cost and decide if it is worth the effort. Standing up for your principles is always admirable, but it is also important to get on with your life.

4

Getting Started

When you were working, your performance spoke for you for 20 years or more. Your boss, peers, subordinates (if you supervised people), customers, clients, patients—they all knew how good you were at what you did.

Now, as a job hunter, you're in a new situation. You have to tell people in writing (the resume) and in person (the interview) how good you are. It's not easy; most of us don't like to brag. And you probably didn't keep years and years of perfect notes of dates, names, and so on. It can be difficult to sit down and write about yourself.

Still, a good inventory of your assets and liabilities is critical to finding that perfect position. Remember, you're not looking for just any job: You're looking for an opportunity suited to your career growth; yes, growth, even at the age of 50 plus. So let's take an easy step-by-step approach toward designing an inventory of your background in outline form, and we can then turn the outline into a resume.

At this point, don't worry about style. Just jot down the facts. And don't worry about which information you'll include in your resume and what you'll omit.

WORK EXPERIENCE FIRST

Start with your work experience. In reverse chronological order (most recent job first), list the employers for whom you have worked, the years of employment (no need to list the month or date), a description of your responsibilities, and your main accomplishments.

I've included a format for you at the end of this section. I suggest you use sheets of notebook paper . . . or you have my permission to make photocopies of this form. Make as many copies as you have had jobs and fill out each one.

Be sure to include military experience, volunteer and community service positions, and every job you've ever held. You can weed out later. Remember: Just because you don't get paid for volunteer work doesn't mean it is worthless on a resume. If you're applying for any kind of supervisory or management position, volunteer and civic experience can show you have leadership, supervisory, and motivational skills.

Even the part-time or entry level jobs you had at the beginning of your career can tell something about you. Did you help out your college roommate in a flower shop over spring break? Work in a department store over Christmas holidays to earn money for gifts? Take a second job to pay your own expenses? These are early indications of a strong sense of responsibility. Did you take an after-school job in high school to pay for a car or college tuition? That shows an ability to set and achieve goals.

You may be wondering why you need to bother writing down early military experience, first jobs, high school jobs, and other early experience, since you won't be including these early jobs in your resume. There are a couple of reasons.

First, taking this inventory prepares you for the interview, as well as the resume because you may remind yourself of some key bit of information (from the early days) that will help in the interview. Take my own example: I worked in my dad's produce market in my teen years. I wouldn't put that in my resume (it was 50 years ago!) because it reveals my age and may prevent me from getting an interview. But that experience taught me the significance of good customer relations, a value that has helped me throughout my career . . . and it's something I might want to mention in an interview.

And even if I don't list the dates of early jobs, I might want to include "previous employment" on my resume (see Part IV for examples).

You'll notice that the "responsibilities" section of the form on page 26 is shorter than the "accomplishments." This is because responsibilities tell an employer what you were supposed to do—accomplishments show what you actually did.

How do you decide what your "key accomplishments" were? It's easy. Just ask yourself: "What did I do in the job that I am proud of? What did I accomplish that other employers would be interested to hear about?" I like to think of it as the "so what?" question.

WORK EXPERIENCE INVENTORY

Years Employed: From ________________ To ________________
Employer ____
____
Your Title ____
Your Responsibilities ____________________________________
____
____
____
____
____
____
Your Key Accomplishments _________________________________
____
____
____
____
____
____
____
____
____
____
____
____
____

This is where so many people write so-so resumes: They fail to focus on their major achievements, which provide the hard evidence that they performed well for other companies and are therefore likely to make a valuable contribution again.

Having trouble thinking of things you have accomplished? Figure 4.1 is a checklist of some thought starters from The Transition Team's publication, *Professional Guide to Career/Life Planning.*

TARGET YOUR INTERESTS

Now take a look at what you enjoy doing, what you do well and what you do less well. Before you actually start writing a resume, you need to have a good idea about what you like in a job and what

Figure 4.1 Checklist of Possible Areas of Accomplishment

- ☐ Trained others (What results? Why were you chosen?)
- ☐ Received awards (For what? Why?)
- ☐ Instituted new procedures (How? Why? What were the results?)
- ☐ Increased sales (How? By how much?)
- ☐ Company used your ideas (Which ones? Why? Results?)
- ☐ Quality-circle accomplishments (Describe)
- ☐ Performed work outside job description (What? How? Why?)
- ☐ Saved the company money (How?)
- ☐ Promotions or upgrading (When and why?)
- ☐ Identified problems that others did not see (How? Results?)
- ☐ Installed new systems (Describe. Why were you chosen?)
- ☐ Increased accuracy (How? With what results?)
- ☐ Decreased scrap (How? With what results?)
- ☐ Reduced inventory levels (How? With what results?)
- ☐ Reduced absenteeism (How?)
- ☐ Increased production (How? What were the results?)
- ☐ Increased efficiency (How?)
- ☐ Reduced manufacturing time (How?)
- ☐ Safety record (Describe)
- ☐ Supervised others (Results)

Source: The Transition Team, *Professional Guide to Career/Life Planning,*

you're capable of doing. We all do best what we like most. And it's important to be clear in your own mind about this—you must have a sharp focus on who you are and what you want to write an effective resume. After all, how can you portray your strengths if you're not sure what they are? Ask yourself the following questions:

- ➤ What are my assets?
- ➤ What are my liabilities?
- ➤ What aspects of past jobs did I like the most?
- ➤ Of which achievements am I most proud?
- ➤ Do I feel more comfortable in a large or small organization?
- ➤ Am I a "people person" or a "loner"?
- ➤ Do I like supervising people?

- ➤ Am I more comfortable with a commission or straight salary?
- ➤ How do I feel about travel?

Your skills and accomplishments are intertwined, and one way to find your strengths is to look at past jobs—which tasks did you like, and which ones did you procrastinate over?

Review the following list of skills to see if this helps you pinpoint the things you're good at and like as well as the ones that are less important to you.

Analysis	Goal setting	Program development
Bookkeeping	Guidance	Research
Composing	Innovating	Scheduling
Computer skills	Inventing	Selling
Creating	Leading	Solving problems
Decisions	Listening	Speaking
Designing	Managing	Supervising
Directing	Motivating	Teaching
Drafting	Negotiating	Troubleshooting
Editing	Organization	Typing/word processing
Evaluation	Planning	Writing

Okay! Once you've outlined your work experience, it's time to look at your educational background. Fill in the following chart:

EDUCATION INVENTORY

College

Degree ____________________

Name of College ____________________

Major ____________________

Special Accomplishments ____________________

____

____

____

High School

Graduated? ____________________

Name of School ____________________

Curriculum ____________________

Special Accomplishments ____________________

____

____

Other Education

____

____

____

____

____

____

CHOOSE YOUR WORDS

Now that you have collected the basic information, you need to start thinking about the words you'll use. When you're writing a resume, think positive, active, and strong. Eliminate any weak, ambiguous, or misleading terms in either job descriptions or job titles. It may help to think in terms of a newspaper headline—you'll never see weak or passive words there.

Words such as "assigned to" or "assisted with," for example, sound wishy-washy and don't add authority to your writing. Cut out most pronouns (I, you, we, they) and articles (the, a, an) to make your writing crisp and to the point. Start your phrases with verbs, and make those phrases short and punchy.

At most, you'll only have 30 seconds to make an initial impression, and if it's a good one, then the reader is likely to look at the entire resume more closely. You want to create the maximum impact with minimum words. Read the following examples of the summary that this vice president wrote as the first paragraph of his resume:

WRONG APPROACH

I have been part of the marketing department for Acme Medical Corp. I report to the president, and am the vice president of marketing. I participate in all strategic business planning. I have contributed to the phenomenal success of the company.

REVISED FIRST DRAFT

Vice president of marketing for Acme Medical Corp. Participated in strategic business planning. Guided company from small entrepreneurial firm to $42 million international leader in arthroscopic medicine.

Winners believe in themselves, and if you're feeling uneasy or apprehensive about your resume, chances are the experienced personnel director will sense it. You can convey a positive attitude and enthusiasm by choosing your words carefully.

When you're writing a resume, choose strong, active verbs that show action and power. The following list of action verbs will do just that; refer to it if you have trouble finding the right word.

ACTION VERBS

achieve
acquire
address
administer
advertise
advise
advocate
affect
amend
analyze
appoint
appraise
approve
arrange
assemble
assess
assume
audit

base
buy
broaden
build

calculate
catalog
centralize
challenge
change
collaborate
collect
conduct
construct
contact
contract
convert
convey
coordinate

decentralize
decrease
define
demonstrate
design
determine
develop
devise
direct
disseminate
distribute
document
draft

edit
eliminate
employ
enforce
encourage
engineer
enlarge
ensure
establish
estimate
evaluate
execute
exercise
exhibit
expand
expedite
expose
extract

facilitate
forecast
formulate
fortify
frame
fulfill

generate
group
guide

handle

identify
implement
improve
improvise
increase
incur
inform
initiate
inspect
inspire
institute
instruct
interpret
interview
introduce

investigate
isolate

judge
justify

launch
lead
liquidate
locate

maintain
manage
market
minimize
moderate
monitor
motivate

negotiate

operate
organize
originate

perceive
perform
pioneer
plan
prepare
present
preside
process

procure
produce
program
promote
prove
provide
publish
purchase

recommend
record
recruit
redesign
reduce
refine
regulate
renegotiate
reorganize
replace
represent
research
reshape
resolve
revamp
revise

select
settle
separate
serve
service

shape
simplify
sell
solve
staff
standardize
stimulate
streamline
strengthen
study
supervise
supply
support
surpass
survey
systemize

test
train
transfer
translate
triple

upgrade
utilize

verify
visualize

write

PART II

WRITING THE RESUME

5

Types of Resumes

Now that you have the elements of your resume outlined, it's time to decide on *format*. In my 35 years in personnel management, I've seen thousands of resumes written in hundreds of formats (some of them pretty weird). I recommend that you leave the very creative formats to the very "creative" types . . . and that you let your experience and maturity speak for themselves.

CHOOSING A FORMAT

After reading all those resumes during my years as personnel manager and as a Transition Team consultant helping job hunters prepare their resumes, I'm convinced that there are only two effective resume formats ("effective" means they will get you an interview):

- *Chronological.* Work history with heavy emphasis placed on accomplishments
- *Functional.* Experience and accomplishments described in general, followed by work history

First I'll discuss the two kinds of resume, and then I'll give you an example of a resume written in both chronological and functional styles. Of the two styles, I prefer the chronological with accomplishments. Most samples in this book are this type.

Chronological

This resume is most often used by applicants who have job experience—which means you! With a chronological resume, a resume writer can highlight achievements and a prospective employer can easily gauge the applicant's experience.

A chronological resume lists the last (or present) job first, and the rest of the job experience appears in reverse chronological order. This also allows you to show the development of your experience, which is often of great interest to an employer.

Here's a brief overview of a chronological resume with accomplishments:

1. Name, address and telephone number
2. Summary
3. Name of the company of most recent employment
   Responsibilities
   Accomplishments
4. Name of the company of next most recent employment
   Responsibilities
   Accomplishments
   (Continue with each successive company going back only 15 years.)
5. Military service
6. Education
7. Accreditations
8. Professional memberships

Functional

Functional resumes are the second most popular resume style. The reason? While they have drawbacks, functional resumes make use of a well-known marketing strategy: They capture the interest of the reader right off. With a functional resume, a writer composes a few brief statements that encapsulate the candidate's primary accomplishments and experience.

The functional resume pays no attention to the chronology of events, but organizes work experience by function, such as "management," "production," "finance," and so on.

This type of resume works best for job hunters with large time gaps between jobs, people who have held many jobs, or those who would like to change careers and need to emphasize their transferable skills. This type of style allows you to stress the scope of your experience without giving dates, and can camouflage past experience.

However, this ability to camouflage past experience is the major drawback of a functional resume: Employers believe it is most often used by candidates who have something to hide. Because of this, many employers become suspicious when a functional resume crosses their desk and focus on finding the flaws.

In addition, a functional resume does not allow you to relate your accomplishments to any one employer or company.

Now let's look at the same resume written in both formats.

CHRONOLOGICAL RESUME SAMPLE

ELLEN S. BLACKSTONE

1234 South Lincoln Roads
Ann Arbor, Michigan 48015
(313) 546-3940

Vice President Finance and Administration with extensive experience in accounting, finance and computer systems.

Demonstrated record of accomplishments with both a Fortune 500 manufacturing company and a large public accounting firm. Financial expertise includes:

Strategic Planning	Financial Reporting
Management Accounting	Operations Analysis
Capital Budgeting	Acquisitions
Computer Modeling	Computer Controls

<u>**PROFESSIONAL EXPERIENCE**</u>:

VICE PRESIDENT FINANCE AND ADMINISTRATION 1989-Present
Twin Star Industries, Ann Arbor, Michigan

Direct finance, accounting and administrative activities for manufacturing operations with annual sales of $150 million.

Responsible for all accounting functions through review and approval of financial information before presentation to the corporate office.

Directly responsible for the supervision of 18 person staff.

- Responsible for decentralization of mainframe computer systems to IBM AS400 system.
- Developed profit center's business strategy resulting in 33% expansion.

SENIOR DIRECTOR ASSET MANAGEMENT - Dallas Processing Center 1986-1989

Managed corporation's major assets including cash transfers, accounts receivable, credit and collection ($33 million), inventories ($68 million) and capital assets and budgets ($600 million).

- In-depth analysis and assessment of accounts receivable resulted in the collection of an additional $2 million.

DIRECTOR OF AUDITING 1982-1986

Managed and coordinated all audit activity for domestic locations, international subsidiaries and corporate office.

- Developed and implemented new and more efficient computer auditing procedures.

Ellen S. Blackstone — page 2

ACCOUNTING MANAGER
KLT Fusion, Inc., Ann Arbor, Michigan — 1976-1982

Joined KLT as Tax Manager. Developed and maintained a centralized control of all taxes for the corporate office and its 25 divisions located throughout the United States. Traveled extensively.

- Analysis of one tax law resulted in $350,000 saving in one transaction.

Previous experience: Cost Analyst, Cost Accountant, Accountant

EDUCATION:

MBA in Finance
University of Detroit

Bachelor of Science
University of Detroit

PROFESSIONAL AFFILIATIONS:

American Institute of Certified Public Accountants

Active member of AICPA Taskforces:
* On-Line Systems
* Audit Impact of Complex Computer Systems

The chronological resume with accomplishments:

- Is easy to read.
- Emphasizes what the over-50 job hunter has to sell: What you did, where you did it, what you accomplished.
- Is direct.
- Replaces objective with clear summary.
- Eliminates superfluous information.
- Uses power phrases rather than windy sentences.

FUNCTIONAL RESUME SAMPLE

ELLEN S. BLACKSTONE
1234 South Lincoln Road
Ann Arbor, Michigan 48015
(313) 546-3940

SUMMARY: MBA degree with extensive experience in accounting, finance and computer systems. Demonstrated record of accomplishments with both a Fortune 500 manufacturing company and a large public accounting firm. Expertise includes: strategic planning, financial reporting, management accounting operations analysis, capital budgeting acquisitions, computer modeling, and computer controls.

CAREER ACCOMPLISHMENTS:

Financial Management

- Directed finance, accounting and administrative activities for manufacturing operations with annual sales of $150 million.
- Managed corporation's major assets including cash transfers, accounts receivable credit and collection ($33 million), inventories ($68 million) and capital assets and budgets ($600 million).
- Responsible for decentralization of mainframe computer systems to IBM AS400 system.
- Developed profit center's business strategy resulting in 33% expansion.
- Analysis of one tax law resulted in $350,000 savings in one transaction.

Cost Accounting

- Developed unique computer model for effectively estimating material costs.
- Active participant in hourly involvement and value analysis groups resulting in marked decrease in cost and increased productivity.
- General ledger reconciliation of Raw Material, Work-in-Process and Finished Goods inventories totaling in excess of $3 million monthly.
- General ledger conversion to on-line computer. Development of Job Cost performance analysis utilizing Multiplan.

General Accounting

- Managed and coordinated all audit activity for domestic locations, international subsidiaries and corporate office.
- Installed new financial reporting packages. Capital investment analysis utilizing Lotus 1-2-3. Account Analysis. Development and implementation of managerial reports, utilizing Multiplan, for auditing and controlling health insurance and utility expenses. Implemented Fixed Asset Management System.
- Developed and implemented new and more cost-efficient computer auditing procedures.

Ellen S. Blackstone page 2

WORK HISTORY:

1982-Present Twin Star Industries, Ann Arbor, Michigan

(1989-present) Vice President Finance and Administration
(1986-1989) Senior Director Asset Management
(1982-1986) Director of Auditing

1976-1982 KLT Fusion, Inc., Ann Arbor, Michigan

Accounting Manager

1975-1976 Detrex Chemical Industries, Inc., Detroit, Michigan

Corporate Tax Accountant

1968-1975 The Barnes Group Inc., Plymouth, Michigan

(1973-1975) Accountant
(1972-1973) Cost Analyst
(1969-1972) Cost Accountant
(1968-1969) Accounting Clerk

EDUCATION:

MBA in Finance - 1973
University of Detroit

Bachelor of Science - 1961
University of Detroit

MEMBERSHIPS:

American Institute of Certified Public Accountants
Active member of AICPA Taskforces:
* On-Line Systems
* Audit Impact of Complex Computer Systems

PERSONAL

Married * Three children * Five grandchildren * Excellent health

References Available Upon Request

The functional resume:

- ➤ Is harder to write.
- ➤ Is harder to read. The recruiter has lots of resumes to read—make it easy for him or her.
- ➤ Looks like you are trying to hide something. The reader wants to know where you worked, what you did.
- ➤ Can be useful if you have an erratic job history (too many jobs in a short time, too many time gaps) or you want to make a significant career change and want to shift the emphasis (example: secretary to customer service rep, production supervisor to human resources manager).

When it comes to choosing resume styles, ask yourself these questions:

- ➤ Do I have good credentials?
- ➤ Do I have a solid work history?
- ➤ Do I meet the educational requirements of this job?
- ➤ Are there no employment gaps?

If you can answer yes to all four of these questions, I recommend that you use the chronological resume with accomplishments. If you answer no at least once, then you might be wiser to choose a functional style.

HOW DOES YOUR RESUME LOOK?

The appearance of your resume—good or bad—can determine whether you get the interview. Preposterous, isn't it? You have more than 20 years' experience and you may be the best in the business, but you can get knocked out of the competition by the appearance of a piece of paper. It may be tough, but that's how the job-hunting game is played.

Look at it from the point of view of the person reading your resume. Sloppy work on the resume may indicate that you are a sloppy worker. A neat, easy-to-read resume suggests that you are a careful, conscientious person.

Some specific tips:

1. Have your resume typewritten, not typeset. The typewritten resume looks like business correspondence written by a

thoughtful, mature person, not a printed circular sent to hundreds of employers.

2. Type the resume, or have it typed, on an electric typewriter or printed out on a laser printer. If you don't have immediate access to this kind of equipment, find a professional service. (Look up "word processing" in the yellow pages.)
3. If boldface or italic type is available, use it for emphasis, but sparingly. Too much special type makes for difficult reading.
4. Make sure that the resume is free from errors: no misspelling, no strikeovers, no use of correction fluid, no erasures, no handwritten corrections. Have someone proofread it.
5. Use white or off-white paper, 24-pound bond, $8^1/_2 \times 11$ inches in size. Never use erasable bond; it is too flimsy and doesn't hold up to handling. Leave colored paper for youngsters who are striving for attention. Attention to *your* resume will come from the content—your superior experience.
6. Use white space and generous margins to make your resume easy to read.
7. Do not include photographs, articles about yourself, reference letters, job performance evaluations, videotapes, audiotapes . . . I could go on and on. I have seen all this "stuff" in resume envelopes and can assure you that resume readers don't have time for the superfluous material.

Now that you understand the two most popular types of resumes, it's time to turn to Chapter 6 and begin creating the kind of resume that will land you an interview immediately.

6

Openers: Identifying Information and Career Summary

Whether you're applying for a job at a Fortune 500 company or a local store, chances are the employers are looking for the same thing: What can you do for their company and for their customers? As they look over your resume, they're going to be asking themselves the same questions:

- ➤ What do you bring to the job?
- ➤ Do you understand the business?
- ➤ Can you communicate well with customers and other employees?
- ➤ Can you demonstrate what you can do?

You must convince the prospective employer that you have what he or she is looking for—and the first part of your resume is designed to do just that.

THE HEADING

Writing your name and address may sound like a snap at first, but there are many things to consider when you write down this basic information. The heading should be clear and easy to read; I don't recommend headings with borders, underlining, or anything else.

Just type it like this:

> **JACK M. WASHINGTON**
>
> 232-15th Street
> Royal Oak, Michigan 48233
> (313) 255-1345

Take note:

- Name typed in capital letters (bold type, if available)
- Double space after name
- Address in initial capital letters and lower case
- Name of state spelled out (not initials—some people don't know that "MI" is Michigan, "MN" is Minnesota, and "MS" is Mississippi)

Omit nicknames and additions (Jr., III, etc.). It's a nice touch to boldface your name, but if you don't have a typewriter or personal computer capable of printing with bold type, the regular font is OK.

If you live on a numbered street, use a hyphen between the street number and the house number. Don't use a post office box number; it makes you look transient. If you don't like your address, you can rent a box service, which can provide what seems to be an exclusive apartment address.

I recommend including only your home phone number. You don't want to risk having prospective employers call you at your present place of work, and a prospective employer may wonder why you feel so free about using your office phone. It may arouse suspicion that your current employer *wants* you to find another job. And I don't like phrases such as "If no answer, leave message at 555-7895." It's just not the kind of arrangement you expect from a mature, experienced person.

Make sure the phone will always be answered. Usually, an employer will be calling several prospective applicants in a day. If nobody answers your phone, or the employer keeps getting a busy signal, eventually he or she will just stop calling. Many employment managers will be willing to call you at home at night, if you ask them to do so in your cover letter.

And speaking of phones . . .

1. Many people over age 50 have teenagers or young adults at home. They love to use the telephone and may even

monopolize it. It's important for you to reach an agreement with your family. Tell them:

I am looking for a job.

I need your cooperation.

You must keep your phone conversations to 5 minutes or less until I find a new job.

The only alternative is to have a second phone installed—not very likely if you're unemployed.

2. Train the family to take your messages. Imagine this conversation:

   JUNIOR: Dad, you had a phone call today.

   DAD: Oh? Who was it?

   JUNIOR: It was some guy about an interview. He said he would call back.

   (Resist the urge to kill. He *is* your son.)

3. Be sure there is a pen and notepad at the phone, reserved for your messages, and be sure that all family members know how to take your messages. A messed-up message can cost you an interview.
4. Install an answering machine. A missed phone call can also cost you an interview. But erase that cute message and replace it with a businesslike one giving your first and last name. If you don't want to buy an answering machine, use an answering service. You can always cancel the service once you have accepted a job.

The easier it is to contact you, the greater the chance you *will* be contacted. So in addition to a phone number, you might consider including a fax line or voice-mail service information.

THE CAREER SUMMARY

"What can this applicant do for me?" That's what an employer is going to be asking, and you must serve up the answer immediately if you want to make sure your resume lands in the "Yes" pile after that first cursory glance.

The best and quickest place to tell an employer what you can do for him or her is the career summary. A good resume is really not much different from an advertisement, and the best advertisers know that to catch a reader's attention, you need a hook—a few lines that are so well-written and sparkling that they guarantee the reader will look at the rest of the ad. You want to "hook" readers and drag them in, *make* them read what you want them to read. Journalists use the same idea when they write a lead, an article's first paragraph. Leads must be short, punchy, and active. If they don't grab a reader right away, they are not doing the job.

In a resume, you hook an employer with a career summary, *not* an "objective." An objective tells the reader what *you* want. Employer's are really interested in what they want. Most objectives are too broad ("A position with potential for advancement in a progressive company") or too narrow: ("A payroll manager position in a company utilizing the ABC computer software system.")

Objectives are considered old-fashioned and out-of-date, which is an impression that job seekers over age 50 want to avoid at all costs. Leave the objective writing to recent graduates. You are a mature, experienced person, ready to make an immediate contribution to your next employer. You want to write a *career summary.*

When you are writing the summary, keep in mind the question that the employer will be asking: "What can this applicant do for me?" Think about your skills, strengths, and accomplishments, and keep it short. Here's an example of what to do, and what *not* to do, when writing a career summary:

WEAK SUMMARY

I am a marketing professional and I have worked for several companies who sell arthroscopic surgical instruments. My company was bought by another company and phased out my job. I live in Boston, and I don't really want to relocate. I've worked for 10 years in marketing and my responsibilities included responsibility for marketing and product management across the country. I got an achievement award two years in a row for my performance.

STRONG SUMMARY

Fortune 500 marketing executive with extensive experience in new product development, marketing management, and strategic planning. Responsible for three new product introductions per year with a 25% annual sales increase.

The weak example is wordy, negative, and includes too much information. Employers are busy and have piles of resumes to plow through; a crisp and clear summary will save everybody time and make everybody a lot happier. If you make an effort to hone and sharpen your career summary, you'll find it will pay off in the end.

Remember: Always accentuate the positive. Don't say what you don't want or can't do, but instead include positive statements that coax the employer to read on. Nobody likes a whiner.

While you want to be concise, be sure to include enough information to give a sense of who you are: Include areas and breadth of expertise, particular strengths, and types of companies for whom you've worked (if it will help sell your abilities).

Read these examples, and then write your own career summary.

HUMAN RESOURCES MANAGER

In-depth experience in all human resource functions. Directed personnel staff of major multi-plant unionized manufacturer. Excellent reputation as developer of progressive programs that served the best interests of the employer and employees.

EXECUTIVE SECRETARY

Administrative assistant to corporate officer. Expert stenographer and word processor. Handled domestic and international travel. Managed executive appointments and meeting schedules.

WELDER

Experienced in all phases of production and job shop welding. Skilled in TIG, MIG, stick, and underwater work.

Now you should have a fairly good sense of your accomplishments and abilities, and how you can use those skills to move on in your career. If you're still having trouble getting it all down on paper, you might find a few sessions with a career counselor helpful.

It's best to put the career summary at the top of the resume, because it helps focus attention on your accomplishments and skills. But I see no need to write a heading for the career summary; there's no purpose in saying "summary" or "career summary." Let the content of the summary speak for itself. It's a brief statement of your qualifications, and it will be followed by the "Experience" section (both professional and educational), which will provide the proof to back up your career summary.

So let's turn to Chapter 7 and learn how to summarize your job experience in a way that's so compelling it will have an employer reaching for the phone before the end of the resume.

7

Describing Your Experience

Regardless of the occupation, experience is the most important asset the over-50 job hunter has. You've lived through more . . . observed more . . . done more . . . than younger workers. You've seen the development of technology, watched more projects go right and go wrong, tested more kinds of activity and observed the results. And you have more experience witn people, too.

HOW TO PRESENT YOUR EXPERIENCE

The idea here is to be forceful without being wordy—so leave out the "I" word. Everybody knows who you're talking about. Begin each sentence with a strong, action verb that carries power and authority. List your professional experience first, in inverse chronological order. Under the heading "Experience," "Professional Experience," or "Professional History," list employer and dates of employment; underneath comes your title and a position summary followed by a bulleted list of entries of significant accomplishments.

Employer

This is the company's name. If the firm is a division of a larger company (often the case today), state it. Describe briefly what the company does, what it makes, what it sells. Be consistent from one company listing to the next, whatever style you choose. Don't take up space with a full company address; the city and state are usually sufficient.

Dates of Employment

The year is generally enough; you can add the month if you want to be more specific. But writing down the exact date (month, day, year) is unnecessary; it looks cluttered and the reader may be distracted from the essence of the resume by having to plow through a day-to-day chronology. Use the words "to present" to indicate your current position.

Title

Don't assume that your job title tells all. Because experience is the most important asset you have to offer a prospective employer, you must be careful to describe your experience in the most effective way.

Here's what I mean:

- Welder. Are you a welder on an automobile assembly line? In a job shop? On a construction site? Different kinds of welding jobs call for a wide range of skills and techniques, depending on the job and type of metals.
- Store Manager. You can manage a neighborhood convenience store, with a predetermined "merchandise mix" and some part-timers working odd hours, or a major department store with broad merchandising autonomy and hundreds of employees.
- Marketing Manager. "Marketing" can mean sales; it can mean market research; it can mean advertising; it can mean combinations of these responsibilities.
- Programmer/Analyst. The title doesn't mean much unless you specify the hardware and software with which you have expertise.
- Plant Manager. This title needs detailed description: How many employees? Is the plant unionized? What product is manufactured? Do you have responsibility for the bottom line?

Statement of Your Responsibilities

Be specific. Your most recent jobs should get the most space. Eliminate pronouns; use action words and "responsible for" phrases. Most companies have job descriptions, and if it's not too complicated or tedious, you can simply write this down to describe your position.

Or you can pick out details from the job description and write your own. Good resources for sample job descriptions are the following publications, which can be found in most libraries:

Dictionary of Occupational Titles
United States Department of Labor, 1977
(with updated supplements)

B.L.R. Encyclopedia of Prewritten Job Descriptions
Business & Legal Reports, Madison, CT, 1986

The Encyclopedia of Managerial Job Descriptions
Business Research Publications, New York, NY, 1986

Developing Job Descriptions
Borgman Associates, Walnut Creek, CA, 1990

The statement of responsibilities should always be followed by a list of accomplishments.

Statements of Accomplishments

This section gives you the opportunity to make your resume stand out. A cold statement of dates, companies, and duties doesn't answer the "so what?" question we raised earlier. And your responsibilities take on more importance when you describe what you did with them.

What sorts of things did you do of which you're especially proud? What did you accomplish for your employer? What did you do that made a difference? Don't just write that you sold machines. Say that sales increased more than 60 percent over the previous year, that you consistently sold 35 percent over your objective.

Take a look at the sample resumes. After the statements of job responsibilities, you'll see phrases like these:

- ➤ Doubled production while reducing work force by 10 percent
- ➤ Increased annual sales from $500,000 to $1.6 million
- ➤ Developed new paperwork procedure resulting in increased proficiency
- ➤ Developed and implemented computerized receiving and shipping records, resulting in 20 percent reduction of error rate
- ➤ Conducted special effort to acquire more mature work force, resulting in 25 percent reduction in employee turnover
- ➤ Wrote procedures manual for secretaries that established uniform clerical procedures for the entire corporation

Whenever possible, your accomplishments should be described in quantitative terms—dollars, percentages, pieces produced, ratings, rankings. But not all accomplishments can be quantified; for some workers, accomplishments will include statements such as "safety committee member," "excellent attendance record," and "trainer of new machine operators."

How Far Back Do You Go?

Scott Byron worked for a large manufacturing firm for more than 30 years when he lost his job following a corporate restructuring drive. But if he listed the number of years he had worked with that company, together with earlier jobs he'd gotten right out of school, his resume would give away his age. He pictured potential employers throwing his resume on the "No" pile strictly because of his age. Fortunately, there's a way around this problem.

If you worked for one company for more than 15 years, list only that company; if you've worked for several companies, give details for no more than 15 years. If you want to indicate more experience (more than 15 years), you can list "previous employment" and give job titles only.

Here's how Scott's resume looked before and after he revamped it to reflect his experience, not his age:

BEFORE

SCOTT W. BYRON
3787 Maryland Street
Memphis, Tennessee 38125
(901) 386-1124 - Work (901) 566-6699 - Home

Solid manufacturing experience with American Foods, Inc.: Logistics, Quality, Production, Customer Service, Personnel, Engineering and Accounting.

PROFESSIONAL EXPERIENCE

American Foods, Inc., (various U.S. locations)

GENERAL MANAGER 1985-Present

Leader of a six-person team commissioned to start-up a manufacturing facility (350+ employees) producing Trim-Down products. Responsible for development, coordination and implementation of accounting and financial functions, manufacturing, engineering, personnel, and quality services.

- Established specifications and selected suppliers for the purchase of all new equipment.
- Completed layout and engineering for the manufacturing operation.
- Designed a pre-blend operation for the sourcing of key ingredients for other co-packers.
- Developed the philosophy, mission and hiring statements for the new facility.
- Developed plans for an analytical and microbiological lab.
- Successfully coordinated tax abatement with County Council.
- Applied for and received local and state training assistance grants from the Indiana Department of Commerce.
- Participated in negotiations for sale of property and equipment between Cross, American Foods, Inc. and Trim-Down.

PLANT MANAGER 1981-1985

Total responsibility for three separate operations: an aseptic refrigerated plant in Indiana, a frozen operation and a dry mix and pack operation, both in Illinois. Accountable to corporate headquarters for scheduling, procurement, production, product quality, safety performance, fixed and variable costs, warehousing, distribution, customer service level, personnel, engineering and systems.

Responsible for 640 UFCW workers and 137 staff and clerical employees. Produced Star Gelatin and Pudding mixes, Ready-To-Eat Puddings, Pudding Stars and Ovenway Stuffing mix.

- Improved the quality rating of finished goods from 75% to 95%.
- Created a JIT production planning and distribution system that increased customer service levels from 94% to 99%+ while reducing finished goods inventory from 37 days to 22 days.
- Managed $1.3 million in cost effectiveness in each of 8 years.
- Reduced workers compensation costs by 67%.
- Created a satellite plant in Mason, Mississippi which utilized the team concept with an all-salaried workforce.

BEFORE

SCOTT W. BYRON PAGE 2

MANUFACTURING AND ENGINEERING MANAGER 1975-1981
Similar responsibilities as above. Managed 320 USW and 55 salaried employees while producing $55 million worth of frozen, concentrated juices. The product line included Squeeze Orange Juice, Orange Aide, Alert and County Fair Lemonade. The plant also processed oranges into concentrate.

- Principal negotiator for the site; negotiated two early contracts without labor interruption.
- Satellited the plant to Denver, Colorado with a savings of $2.4 million.
- Reduced spoilage costs by 45%.
- Institutionalized good manufacturing practices and significantly improved housekeeping.

BUILDING SUPERINTENDENT - SQUEEZE 1971-1975
Managed manufacturing and maintenance for Squeeze with $50 million annual manufacturing costs utilizing 120 UFCW workers and a staff of nine.

- Introduced new flavors and operating standards.
- Extended the incentive system to 110% of standard, saving over $1.5 million annually in labor costs.

BUILDING SUPERINTENDENT - OAT FLAKES 1968-1971
Implemented new operating schedule that increased yields from 78% to 85% while stabilizing employment.

OPERATIONS FOREMAN 1964-1968
Direct and schedule the activities of 56 production workers. Responsible for continuous work-flow and overall quality control. Maintained production records and inspected products during and after production to insure conformance to company specification. Additional duties included: enforcing safety rules and demonstrating time-saving and labor-saving techniques. Responsible for making recommendations on wage increases, awards and promotions. Also handle disciplinary problems and resolve grievances.

THIRD SHIFT FOREMAN 1961-1964
Direct and schedule the activities of 32 warehouse workers engaged in the loading and unloading of products and raw material. Responsible for continuous work-flow and overall quality control. Maintained records and inspected and reviewed shipping documents. Additional duties include enforcing safety rules and demonstrating time-saving and labor-saving techniques.

EDUCATION:

M.B.A. Marketing
Eastern Illinois University

B.S. Industrial Management
Mississippi State University

INTERESTS & MEMBERSHIPS:

United Way Captain, Day Care Center Board President
High School Booster Club President, Soccer Association President, Coach and Referee, JA Board Member, High School Youth Group Leader, Little League Coach

AFTER

SCOTT W. BYRON
3787 Maryland Street
Memphis, Tennessee 38125
(901) 386-1124 - Work (901) 566-6699 - Home

Solid manufacturing experience with American Foods, Inc.: Logistics, Quality, Production, Customer Service, Personnel, Engineering and Accounting.

PROFESSIONAL EXPERIENCE
American Foods, Inc., (various U.S. locations)

GENERAL MANAGER 1985-Present
Leader of a six-person team commissioned to start-up a manufacturing facility (350+ employees) producing Trim-Down products. Responsible for development, coordination and implementation of accounting and financial functions, manufacturing, engineering, personnel, and quality services.

- Established specifications and selected suppliers for the purchase of all new equipment.
- Completed layout and engineering for the manufacturing operation.
- Designed a pre-blend operation for the sourcing of key ingredients for other co-packers.
- Developed the philosophy, mission and hiring statements for the new facility.
- Developed plans for an analytical and microbiological lab.
- Successfully coordinated tax abatement with County Council.
- Applied for and received local and state training assistance grants from the Indiana Department of Commerce.
- Participated in negotiations for sale of property and equipment between Cross, American Foods, Inc. and Trim-Down.

PLANT MANAGER 1981-1985
Total responsibility for three separate operations: an aseptic refrigerated plant in Indiana, a frozen operation and a dry mix and pack operation, both in Illinois. Accountable to corporate headquarters for scheduling, procurement, production, product quality, safety performance, fixed and variable costs, warehousing, distribution, customer service level, personnel, engineering and systems.

Responsible for 640 UFCW workers and 137 staff and clerical employees. Produced Star Gelatin and Pudding mixes, Ready-To-Eat Puddings, Pudding Stars and Ovenway Stuffing mix.

- Improved the quality rating of finished goods from 75% to 95%.
- Created a JIT production planning and distribution system that increased customer service levels from 94% to 99%+ while reducing finished goods inventory from 37 days to 22 days.
- Managed $1.3 million in cost effectiveness in each of 8 years.
- Reduced workers compensation costs by 67%.
- Created a satellite plant in Mason, Mississippi which utilized the team concept with an all-salaried workforce.

AFTER

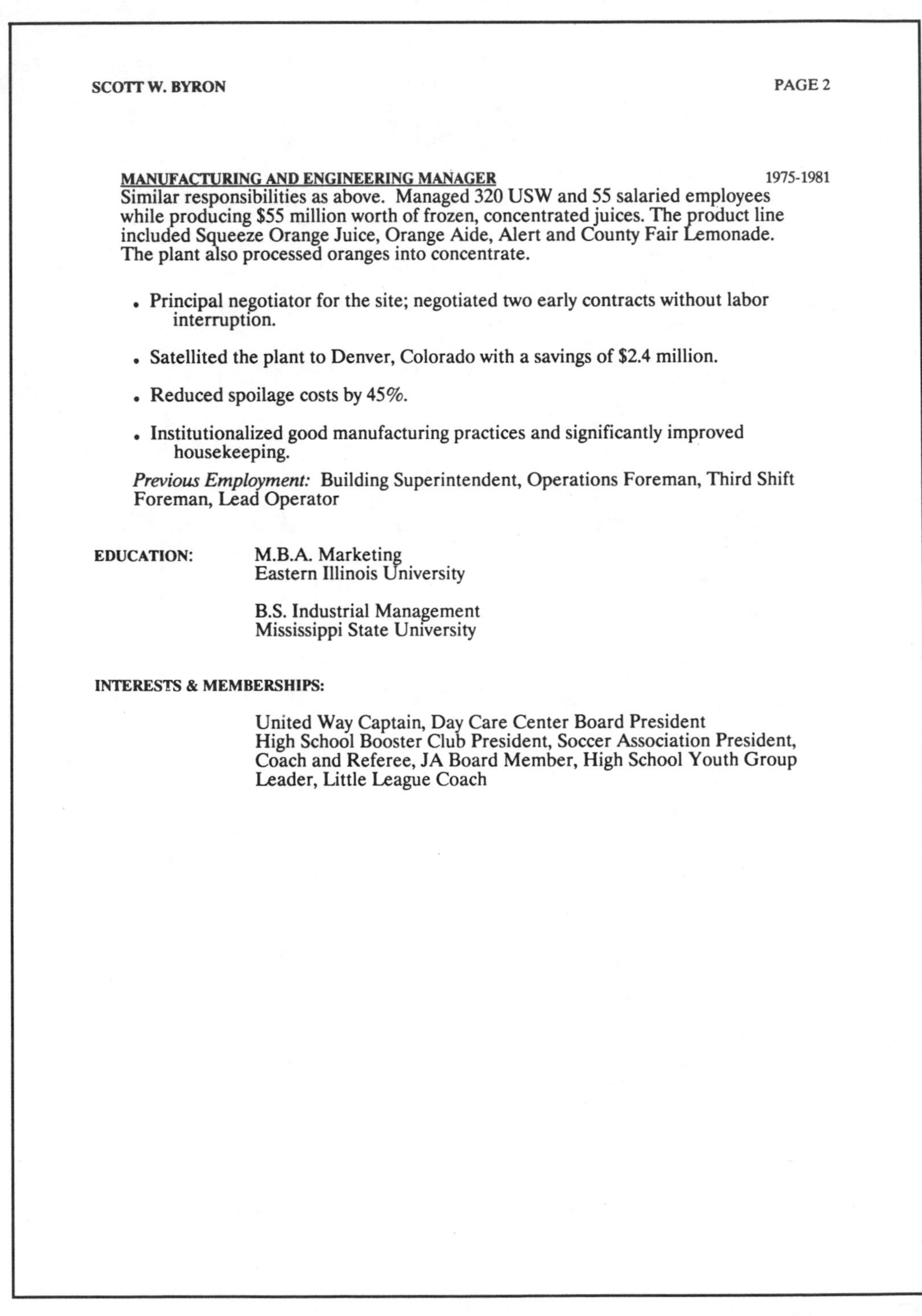

SCOTT W. BYRON **PAGE 2**

MANUFACTURING AND ENGINEERING MANAGER 1975-1981
Similar responsibilities as above. Managed 320 USW and 55 salaried employees while producing $55 million worth of frozen, concentrated juices. The product line included Squeeze Orange Juice, Orange Aide, Alert and County Fair Lemonade. The plant also processed oranges into concentrate.

- Principal negotiator for the site; negotiated two early contracts without labor interruption.
- Satellited the plant to Denver, Colorado with a savings of $2.4 million.
- Reduced spoilage costs by 45%.
- Institutionalized good manufacturing practices and significantly improved housekeeping.

Previous Employment: Building Superintendent, Operations Foreman, Third Shift Foreman, Lead Operator

EDUCATION:

M.B.A. Marketing
Eastern Illinois University

B.S. Industrial Management
Mississippi State University

INTERESTS & MEMBERSHIPS:

United Way Captain, Day Care Center Board President
High School Booster Club President, Soccer Association President, Coach and Referee, JA Board Member, High School Youth Group Leader, Little League Coach

Reason for Leaving

And a word about including "reasons for leaving" your last job—don't. The only time to include that information is if the departure was due to a plant closing (for example: "Available due to plant closing"). It's best to include only information that will help get the interview.

OTHER EXPERIENCE

Awards and Associations

If you belong to trade associations or professional organizations, put them down. For some jobs, it's almost a necessity to belong to some of these organizations, and they suggest you're interested in furthering your professional activities.

Also list any awards or honors you received. If you have quite a few, it is better not to try to squeeze them in. Instead, write "See addendum" and attach a separate list. This also applies if you've published many articles and publications.

Military Service

There is definitely a place for your service in the armed forces with an honorable discharge, although you should keep it brief. Exception: you won't want to be so brief if your military career has been a major part of your background, and you're seeking a civilian job after retirement from the armed services. In this case, you want to stress your military experience that is most relevant to the kind of work you're interested in getting as a civilian.

Community Activities

These days, more and more companies are brushing up their public image by participation in community activities. They may look favorably on employee participation in fund drives, board memberships, and so on. A potential employee who is active in community projects comes across as a caring, responsible person with broad interests and plenty of get-up-and-go.

Accreditations and Licenses

Any credential related to your job should be included in your resume: licensed engineers, real estate brokers, beauticians, CPAs, and so on.

Patents and Publications

Patents are particularly important to engineers and scientists, and publications are especially helpful for teachers, consultants, and lawyers. If you have a patent or have been published, list it.

WHAT NOT TO INCLUDE

Leave your family at home. You don't need to mention that you're married and have children aged 18 and 24—it's a mid-life giveaway. Should the question come up during the interview (and despite its illegality, many employers still ask people, especially women, if they are married or have children), you'll be able to point out the benefits of having an older family. Point out there's no danger of sick children to leave work for, day care problems, employer-paid maternity benefits, or maternity leave.

But if you include personal information in your resume first, you may not be there to point out the pluses in this situation.

Extracurricular Activities

Some extracurricular activities can be impressive and should be included, especially those that indicate you are an active, high-vitality sort of person (e.g., a Little League coach, officer in a well-known charitable or community organization). In most cases, however, you've attained enough educational and job-related experience to make these extra activities appear superficial and it is better to omit them.

Birthdate

This may be obvious, but you don't want to give any ammunition to an employer who may be prejudiced against the 50+ job seeker. Leave your birthdate out.

Race, Religion

These details are unrelated to job performance and should not be included on any resume.

Height, Weight

Unless you're looking for a job as a professional wrestler, these items have nothing to do with job performance.

Social Security Number

Keep in mind the purpose of a resume: to help you get an interview, not to reveal as much information as possible.

References

I don't recommend any mention of references on a resume. Why?

- They take up space better utilized for stressing experience and accomplishments.
- You don't want your references called by strangers who received your resume. References are valuable; they have to be prepped to receive those calls.

If you need a filler . . . or if you feel strongly that you must say *something* about references, say, "References available upon request."

I do recommend that you have a list of references with you when you appear for an interview. When you are asked about references, you can just hand over your typewritten list—with names, titles, addresses, and phone numbers. This says something about you, that you came prepared, that you are organized.

Here's an all-too-typical scenario:

INTERVIEWER: Do you have references?

CANDIDATE: Yes.

INTERVIEWER: Please give me the names and contact information.

CANDIDATE: Oh! I have them at home. [or] May I borrow your phone book so I can look them up?

Here's how it *should* go:

> INTERVIEWER: Do you have references?
>
> CANDIDATE: Yes. Here they are. [Hand over your list, neatly typed on the same stationery as your resume.]

Here's a sample of a well-prepared reference list:

Ellen R. Kensington

References

Noah D. Johnson — 212-541-0230
President
American Publishing Company
3637 Blackstone Ave.
New York, New York 10010

Julia C. Anderson — 718-649-1300
Director of Human Resources
Brooklyn Home Products, Inc.
716 Beverly Road
Brooklyn, New York 13451

James L. Maxwell — 609-456-9870
Production Manager
Terminal Manufacturing Co.
423 Hilltop Drive
Trenton, New Jersey 34628

Roberta R. Murray — 718-634-3243
Mayor
City of Oak Park
Oak Park, New York 15433

Salary Information

My advice is to leave out salary information, even when an ad specifically asks for it. You want to focus the reader's attention on your in-depth experience and on the value you bring to the job, not on the cost.

Salary may also keep you from getting the interview. If it's "too high" you won't get in the door. If it's "too low" the employer may conclude that you don't have the skills or experience.

Yes, you *are* taking a chance; an employer may put resumes without salary information on the "No" pile. But a well-crafted resume will make you look like such a well-qualified candidate that the boss will say: "I want to meet this person!"

Leave salary discussions for the last item in your interview. Let your maturity, your experience, your accomplishments make you such an attractive candidate that salary negotiations become easy.

8

Education

Your education provided you with a good start in life, but keep in mind that your most important assets are the experience you have gained and your accomplishments during the past 25 or more years.

That's why you should emphasize your experience and maturity by giving top billing to experience. List your education last. (The young graduate with little experience will list educational achievements at the top of the resume.)

An exception: In some professions, educational credentials are an absolute must (for example, physicians or college professors). In these cases, place education at the top of the resume.

LISTING COLLEGE CREDENTIALS

The name of the institution and the degree awarded is all you need to go into: You don't have to say you earned your degree in 1960. This should be enough:

> B.S., Electrical Engineering, Pennsylvania State University
>
> M.S., Electrical Engineering, Massachusetts Institute of Technology

However, if you earned a degree within the past 10 years, you can include the date. This shows you have enough gumption to go out and improve your educational credentials and keep your career current.

Any academic degree or even attendance at college means you automatically can eliminate mention of your elementary and high school experience unless you went to a prestigious prep school. If you didn't go to college, then mention that you're a high school graduate. But don't apologize for not going to college or having all the qualifications the employer has advertised; stress only the positive.

Class standing doesn't matter unless it's very high, but honors, scholarships, and awards all merit a mention.

So if you have a college degree:

- Omit information about high school.
- List the highest degree first.
- Give either the full name of the degree or initials. If you use initials, punctuate with periods (see examples).
- Omit the date of the degree.
- Give the full name of the college, but not the address.
- Resist the temptation to give too much detail. Include just the name of the degree (include the major if you think it will help) and honors (if you think they are significant). Omit your grade point average; after all, it was years ago.

OTHER EDUCATIONAL CREDENTIALS

Not everyone has degrees, but that lack doesn't necessarily have to stand in your way to career success.

> Candidates must possess a B.S. in finance or accounting.
>
> Should have a minimum of one year managerial experience and a degree in accounting.
>
> College degree required.
>
> The ideal candidate will have a marketing degree.
>
> Position requires technical degree plus MBA.

These are all quotes from classified ads. (Ads, incidentally, are not the best resource for job hunters.) As I read these ads, I thought about the successful people, without degrees, whom I know. For example:

- The owner of a nationwide consulting firm
- The manager of a building components plant
- The president of a major store chain (he has a degree—but in pharmacy, not business management)

- The personnel director of a chain of supermarkets
- A consultant who conducts seminars all over the country

Employers, especially personnel and human resources managers, have a tendency to say that the vacant position requires a degree, even when the work doesn't really require it. If you have a degree, great. Your formal education may help you get the interview.

If you don't have the educational "requirements," the *experience* section of your resume will have to be very persuasive. You must convince the employer that you can do the job, regardless of your formal educational background.

What to Include

If you didn't graduate from high school, it's best to omit the "education" section from the resume. Why raise the question in the resume reader's mind, even before he or she meets you? You and I know that lots of us quit school in our teens to help on the farm, help in Dad's business, or to go into the military. But the youngish personnel manager might not understand that.

What about the GED? All states and most employers recognize the GED as high school equivalent. I recommend, if you have earned your GED, that you say "high school graduate" on your resume. And you don't need to feel that you might be second rate compared with a traditional high school graduate. According to the American Council on Education, GED grads often have higher literacy scores than those who received their high school diplomas the more conventional way. In recent council surveys, three out of five employed GED-holders held jobs that require high degrees of literacy—positions in sales, administrative support, and technical occupations.

What about Company Training?

Make a list of the training provided by your employers. Then select the training, for your resume, that will enhance your chances of getting the interview.

Kinds of training to include:

- Apprenticeship
- Technical courses: statistical process control (SPC); blueprint reading; computer-aided design (CAD)

Kinds of training that may have had value but don't add much to your credentials include human relations, short courses on "how to be a better supervisor," and telephone techniques.

What about Adult Education Courses?

You may want to include courses on computer skills, financial controls, and technical subjects. Omit the cake decorating, square dancing, and poetry courses.

RETURNING TO SCHOOL

Too old to learn? Don't you believe it! No matter what your specialty, it's a good idea to keep up with the current knowledge and skills of your profession.

If you don't have a high school diploma (and even if you do) you might consider getting some extra training to add some pizazz to your resume. If you have thoughts about going back for more education, you're not alone. Older adults are returning to the classroom in droves, and colleges and universities are seeking out older students with or without high school diplomas.

Today, institutions are offering older students:

- Specific counseling programs
- Adult-oriented courses
- The College Level Examination Program (CLEP) (credit for life and work experience)
- Scholarships or subsidies for senior citizens

But colleges aren't the only choice if you're considering going back to school for more training. Community adult education classes offer job-directed training in a wide range of areas, including computer programming, TV repair, accounting, and real estate. Correspondence schools offer independent study by mail. And don't forget the community colleges and TV courses—check them out!

PART III

YOU HAVE THE RESUME—WHAT DO YOU DO NOW?

9

Cover Letters and Thank-You's

By now you've studied this book and come up with a well-crafted resume, but that resume alone isn't enough. You need something to personalize that resume, to tailor it to the specifics of the position for which you're applying. You need a cover letter that is just as carefully written as the resume it accompanies.

You'll want to use a cover letter when you're answering an ad, following up after a network referral, or after making a cold call. So just what should your cover letter do?

- ➤ Make your introduction for you
- ➤ Stir up some interest from the employer, as a preview of the main feature—your resume
- ➤ Help tailor the information in your resume for the specifics of the job

Cover letters aren't difficult to write if you break them down into three separate parts, and tackle them individually. Basically, the three parts in every cover letter are the opening, the main body, and the closing. With a little forethought, you can create a strong, favorable impression in a few paragraphs that will provide just the right entree for your resume.

THE OPENING

State your purpose in writing. Are you answering an ad? Writing as a result of a suggestion from a mutual acquaintance? Or are you sending along a resume as a result of research that stimulated your interest? Don't waste words: Let the employer know right away why you're writing.

The following example illustrates what I mean:

> Dear Mr. Jones:
>
> In response to the advertisement in the Lancaster *New Era,* I am applying for the position of Public Relations Associate.

THE MAIN BODY

Now you'll want to include a paragraph or two discussing what you have to "sell" to the employer: your unique skills and experiences. Try to relate these to the needs of the prospective employer. Of course, this will take some research on your part, but you'll write a better cover letter if you can refer to the employer's business needs and the problems confronting the industry. This section can really put the sizzle in your sell.

> My qualifications, as shown in the enclosed resume, match the described position. I have in-depth experience in the field of hospital public relations, and I enjoy a busy environment with plenty of challenges. My diverse experience in fundraising, publication production and design, speech-writing, and budget preparation could be valuable in these days of fiscal restraint.

THE CLOSING

Here's where you make your action statement. Of course, what you want is the interview. You should ask for the appointment or even say that *you* will take the next step, that you will follow up. Remember to include a thank-you sentence.

> I would appreciate an opportunity to meet with you personally and explore future employment with General Hospital. I'll be in touch with you at the end of the week to arrange a brief meeting. Thank you for your time and consideration.

If at all possible, address your letter to a specific individual. Letters addressed to the "Personnel Department" get little attention—there's just too much competition. Be sure to tailor your letter to the recipient's needs, and be certain that your grammar, spelling, and punctuation are perfect. And always use good stationery; match the paper used on your resume. Note: The examples in this chapter would be prepared, as any business correspondence, on an 8 1/2 by 11 sheet of stationary with appropriate margin. Our examples are not shown full size.

As we discussed earlier, most jobs are not filled by answering ads, but you have nothing to lose, and everything to gain, by responding to the classifieds. Here's a sample letter showing a good way to do just that.

LETTER IN RESPONSE TO AD

Charlotte S. Smith
104 S. Oak Street
Columbus, Ohio 32145
(501) 645-7856

November 15, 1991

Mr. James W. Baker
P.O. Box 456
Columbus, Ohio 32145

Dear Mr. Baker,

In response to the advertisement in the Columbus Star, November 15, 1991, I am applying for the position of Executive Secretary.

My qualifications, as detailed in my enclosed resume, match the described position very well. I have over 20 years' experience in the field, type 75 wpm, and take shorthand. Supervising and organizing a well-run office are my strengths. I enjoy a busy work environment and love the challenge of learning new things. I've recently added several new software packages to my portfolio of skills.

I am available for an interview at your convenience and would enjoy the opportunity to discuss this position and my qualifications in more detail. I look forward to hearing from you.

Sincerely,

Susan Smith

As I mentioned earlier, it also pays to make "cold calls" to individuals who may be in a position to hire you, whether now or sometime in the near future. After making such a call, you would send a letter like the following sample, along with a copy of your resume.

LETTER FOLLOWING A "COLD CALL"

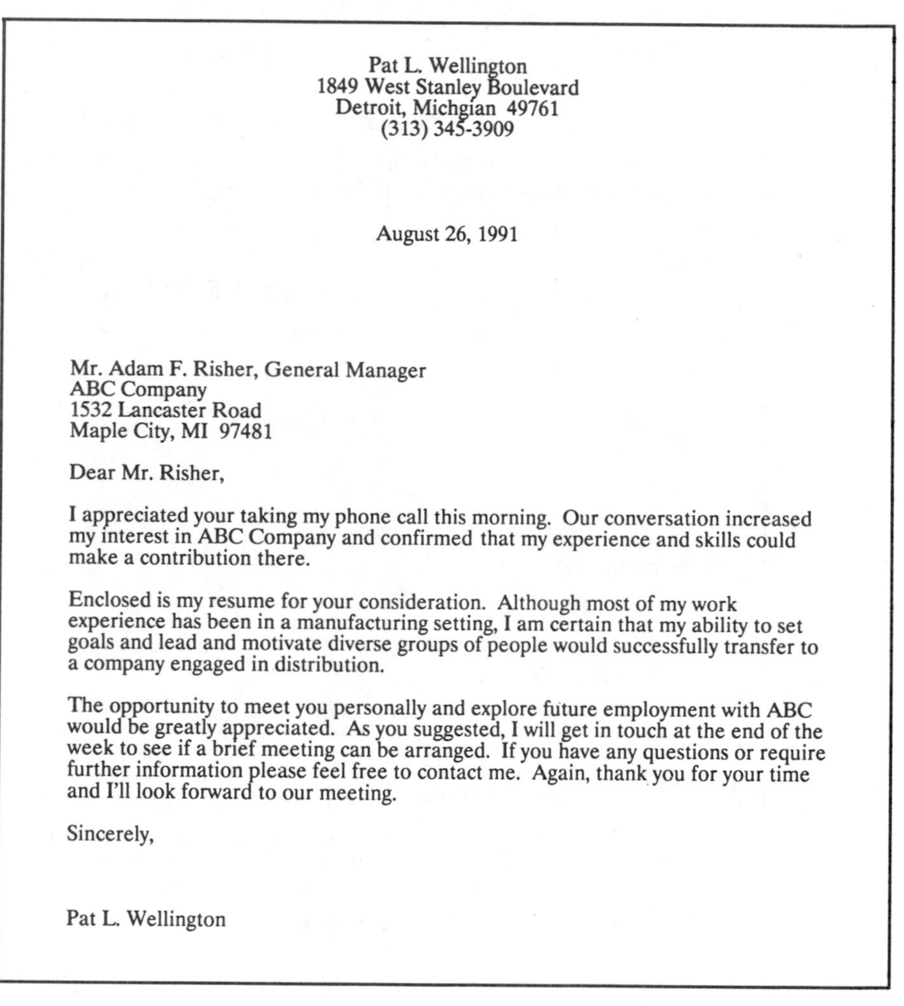

Pat L. Wellington
1849 West Stanley Boulevard
Detroit, Michgian 49761
(313) 345-3909

August 26, 1991

Mr. Adam F. Risher, General Manager
ABC Company
1532 Lancaster Road
Maple City, MI 97481

Dear Mr. Risher,

I appreciated your taking my phone call this morning. Our conversation increased my interest in ABC Company and confirmed that my experience and skills could make a contribution there.

Enclosed is my resume for your consideration. Although most of my work experience has been in a manufacturing setting, I am certain that my ability to set goals and lead and motivate diverse groups of people would successfully transfer to a company engaged in distribution.

The opportunity to meet you personally and explore future employment with ABC would be greatly appreciated. As you suggested, I will get in touch at the end of the week to see if a brief meeting can be arranged. If you have any questions or require further information please feel free to contact me. Again, thank you for your time and I'll look forward to our meeting.

Sincerely,

Pat L. Wellington

Sometimes, nothing can improve on the direct approach, where you just sit down and honestly explain your job situation and that you would like to be considered for a job. This letter, of course, would also accompany a copy of your resume.

DIRECT APPROACH LETTER

September 12, 1990

Jean L. Murphy
102 S. Green St.
Pleasanton, MI 46205
(313) 232-9034

Mr. Joseph R. Roxner
Director of Marketing
Luxon, Inc.
Box 45
Pleasanton, MI 46205

Dear Mr. Roxner,

The plant where I have been employed as a General Sales Manager for the past 12 years will be closing this month. I am very interested in being considered for a position in your plant either in Sales, Marketing or in any capacity my skills could be used. I am very familiar with all phases of the manufacturing of truck parts.

Enclosed is my resume for your review. My qualifications also include an outstanding sales record and commendations by supervisors for the quality of my work and my ability to motivate others.

I have over 20 years management experience in the manufacturing arena and look forward to an opportunity to use my knowledge in a new setting. I would like to meet with you briefly to discuss career opportunities at Luxon. I will call next week to see if your schedule permits such a meeting. Thank you for your consideration.

Sincerely,

Jean L. Munn

If you're retiring and are interested in doing some consultant work, a retirement letter can fill the bill; this is the way to announce your retirement and let prospective employers, and people in your network, know you are available. Here's an example of a good letter written by a retiring executive to a colleague, letting him know in a warm, friendly, yet professional way that she has voluntarily retired and is now available for new challenges.

NETWORKING LETTER ANNOUNCING RETIREMENT

Alice B. Carroll
218 Northlawn Ave.
Athens, Ohio 57487
(402) 478-49032

June 13, 1991

Mr Robert F. Wellington
Creative Services
314 W. Oak Street, Suite 203
Dayton, Ohio 48938

Dear Bob,

I wanted to let you be the first to know that I decided to accept the voluntary early retirement package offered by Belview Systems as of July, 1991. However, and most importantly, I do not plan to sit at home and rock for at least another seven to ten years. You know how I love to keep busy and productive. Consequently, I have written a resume in order to begin my job search and am forwarding a copy to you and several other friends.

Should you become aware of any of your friends, business associates, etc., who may be in the market for someone with my skills and capabilities, I'd appreciate a call. Actually, any assistance or advice from you would be helpful. I've always valued your unique perspective on things.

I'll give you a call in the near future and let you know how my search goes. Maybe we can get together. Thanks in advance for your interest.

Very best regards,

Alice

In a slightly different vein, here's an example of how a personnel executive used a networking letter: By dropping a mutually familiar name, he's getting a foot in the door and sending a strong message of professionalism and competence along with his resume.

NETWORKING LETTER TO MUTUAL FRIEND

2397 Coolidge Road
Woodside, CA 91321
June 30, 1992

Mr. Charles C. Arbor
President
MUTUAL DISTRIBUTORS, INC.
1313 Broad Street
Princeton, CA 91333

Dear Mr. Arbor:

When I mentioned to Mike Sebo that I was looking for a new position, he suggested I contact you. Mike said that you have a vast network in the food distribution industry and that you might be willing to meet with me to discuss career opportunities.

My extensive human resources administration background includes in-depth experience in:

- *recruiting*: hired employees at all levels, order selectors to vice presidents
- *benefits*: designed flexible program which stabilized costs
- *compensation*: developed pay-for-performance plans at all levels
- *information systems*: directed the in-house development of HRIS

I am enclosing a copy of my resume. I'll call you later this week to see if we can set up an appointment.

Very truly yours,

Terrance G. Oswald

Enc.

And here's an example of an excellent letter sent to recruiters showcasing this job hunter's resume and experience.

NETWORKING LETTER TO RECRUITERS

May 21, 1991

Joseph P. Hammer
Technical Placement Associates
4000 West Elm Street
Troy, MI 48084

Dear Mr. Hammer,

As a result of the current downsizing at Americo, Inc., I am planning a career move and would like assistance from you and your firm to make this a successful and exciting transition.

As you can see by the enclosed resume, I have extensive experience in Sales Management, Marketing and Strategic Planning. I'm viewed as an innovate leader with excellent organizational and communication skills. In addition, I have the financial experience to manage and analyze budgets in excess of $6MM.

Please review my qualifications against the requirements of your current clients. I am happy to consider a relocation for the "right" position. If you should need further information, feel free to call me either at my office 647-9304 or home 829-3948. I will touch base with you in the next week or so. Thanks again for your help.

Very truly yours,

Robert B. Fletcher

THANK-YOU LETTER

It may sound old-fashioned, but a thank-you letter is always a good idea after an interview. You've been around long enough to remember when "please" and "thank you" were standard practice: You know how important those words can be. In my 35 years of recruiting, I don't think I've received thank-you letters from more than 10 percent of the people I've interviewed. Why bother to do it?

- It's another chance to get your name in front of a prospective employer.
- So few people do it, you'll stand out.
- It's an opportunity to include information you forgot ("I meant to mention that I have 10 years of experience supervising the company's apprenticeship program").

When do you send thank-you notes? Immediately after each interview or contact. Don't neglect to send one to the people who gave you leads, advice, and information. Thank-you notes for job referrals are also common courtesy and a good way to announce your new position. Handwritten thank-you notes are acceptable, especially if you want to make them seem more personal.

Here's a sample thank-you note that was sent after an interview.

THANK YOU FOR INTERVIEW

Wesley P. Harding
5324 Oak St.
Maywood, Illinois 37845
(213) 475-4950

August 16, 1991

Mr. Sherwood P. Anderson
Ex-tech-Co
3728 North Washington
Chicago, IL 37825

Dear Mr. Anderson,

I sincerely enjoyed our meeting today and wish to thank you for your insights and the information provided about opportunities at Ex-Tech-Co. The tour was fascinating and your staff impressed me with their enthusiasm and openness.

After reviewing our discussion, I am convinced that there is a solid match between my expertise and your needs. The position you described would indeed be the challenge I've been seeking. I'd like to propose a follow up to our meeting. After today, I know that the demands on your time are very great, so I will make myself available at your convenience.

Again, thank you for your time, and I'll look forward to hearing from you in the near future.

Sincerely,

Wesley Harding

Here are some more sample letters that will help you get started:

Letter Using Network Contact
Networking Letter to Friends
Thank-You for Referral
Thank-You for Referral—Job Announcement

LETTER USING NETWORK CONTACT

April 11, 1991

Dana P. Martin
431 W. 102nd Avenue
Rolling Hills, California 95088
(902) 786-9089

Ms. Wilma G. Godwin
Ameritech, Inc.
701 W. 27th Street
Modesto, CA 95351

Dear Ms. Godwin,

Recently, I learned through Bill Kennedy, your General Accounting Supervisor, of plans to expand the facility at Modesto. I am interested in a position involving Industrial Relations or Personnel Management.

I have over twenty years' experience and a broad knowledge of all phases of plant management. I also worked on time study and job classification projects, which resulted in a 15% increase in production efficiency. The plant I am currently employed by will be closing soon, so I will be available at the end of the month.

My resume is enclosed for your review. I would appreciate a personal interview to explore opportunities within your company. My plans include a trip to your area next week, so I will call ahead to set up a meeting time then. Thank you for your consideration.

Sincerely,

Dana Martin

NETWORKING LETTER TO FRIENDS

Alexander M Lawton
2144 Northlawn
Middleton, CT 02819
(202) 278-3902

December 19, 1991

Dear Howard,

I am contacting several friends to ask for assistance and advice. You may have heard about the merger here at XYZ company. We've been bought out by Comco, Inc. and our facility will close at the end of July. As a result, I find myself in the job market.

After a lot of soul searching and reviewing my career goals, I've decided to seek a managerial position in the accounting/finance/MIS field. I'd prefer a company that is growing and expanding, but I am open to almost any good opportunity in any industry.

I know that the positions I am looking for usually do not appear in the classifieds, so I am setting up my own network of personal contacts. I am hopeful that you may be able to refer me to a friend, acquaintance or business associate, or ideally an individual in a company who would be interested in my capabilities. The enclosed resume will bring you up to date on my career and give you a good idea of the type of qualifications I possess.

I appreciate your interest and look forward to talking with you in the near future.

Regards,

Alex

THANK-YOU FOR REFERRAL

Jeanne A. West
12456 Windwood Lanes
Beverly Hills, Michigan 48008
(313) 632-5893

July 17, 1991

Linda A. Holland
3894 Sutton Place
Southfield, MI 49784

Dear Linda,

Thank you so much for referring me to Sam Harrison at Pioneer, Inc. We had a very interesting discussion over the telephone and he asked me to forward my resume. Although they have no openings at this time, he was very enthusiastic about my qualifications. I emphasized the work I did at Telecom, as you suggested and that proved very helpful. He suggested I stay in touch with him as they are approving new budgets in the next few weeks.

I'm still trying to set up a meeting with Herb Scott, but he's been out of town. His secretary remembered you, however, and I think that will help me to get an appointment.

Again, thank you so much. I will continue to keep you posted on my progress. With your assistance and support, this job search is bound to be a success.

Regards,

Jeanne

THANK-YOU FOR REFERRAL—JOB ANNOUNCEMENT

Dear Warren,

Happy New Year!! Thanks to your referral and kind advice, I am pleased to announce my new position with XYZ Company as their Director of Marketing. I can't tell you how much your support has meant to me. A job search can be a lonely endeavor, if one doesn't have a solid network of friends like you. I'm truly grateful.

Now that I'm re-employed, we'll have to have lunch on me. I'll be calling you soon from my new office. Again, thanks, and if I can ever be of any assistance to you, please let me repay the favor.

Regards,

10

Using Your Resume

Diane Landers was a mid-level manager whose company cut back its job force leaving Diane and two friends out of work. One of Diane's friends placed her faith in a private placement agency, and went out on two interviews in a month. Diane's other friend spent his time answering newspaper ads, and sent out three or four resumes weekly. While Diane used both these strategies, she also started networking: She contacted friends, neighbors, family members to ask about jobs—but she didn't stop there. She enlisted the aid of her accountant, minister, banker, lawyer, classmates, school friends, her hairdresser, and dry cleaner. Then she went through her Rolodex and contacted all those names, too. Within a month, Diane had secured 10 interviews and landed a job better than her last one. Her two friends, who were still looking for work, congratulated Diane for her "luck."

Luck had nothing to do with it. All three had good experience and solid resumes, but what Diane *did* with hers made all the difference. You can have the best resume in the world, but it won't matter if it sits in a file drawer and nobody ever gets to see it.

If you've followed my advice in the earlier chapters, you have a terrific resume that should help you land an interview. Now, the first thing you must do with this resume is to get it into the hands of as many contacts as possible.

NETWORKING

The best way to circulate a resume is by networking. In fact, 7 out of 10 people who are looking for a job find one this way—by talking to friends, friends-of-friends, acquaintances, and business contacts. Networking is just a way of getting somebody you know to introduce you to somebody you've never met; once that's done, you're on your own in selling your skills and ability.

Although an effective job seeker needs a well-written cover letter and an excellent resume, a good network can find and open doors when there doesn't appear to be a job opening anywhere in sight. There is, however, a right way and a wrong way to go about utilizing a network, and in the next few pages, I'll show you the difference.

First of all, you don't need a bulging Rolodex or a huge circle of acquaintances to make networking work for you. All you need to start with is a list of 25 people who are most likely to be able to help you.

You don't have that many contacts? Of course you do. Read over the following list, and then sit down and fill in as many names as you can. If you write down only one name per category, you will have 25 contacts to call right now, who may know 25 other contacts, who may know another 25, and on and on. See how it works? Here's the list:

1. Friends
2. Community members in clubs/associations
3. Suppliers or previous customers
4. Immediate family members
5. Aunts/uncles/cousins
6. Editors of trade journals
7. Step-relatives
8. College friends/faculty
9. Participants in professional meetings you've attended
10. Your mail carrier
11. Your dry cleaner
12. Local politicians
13. State/federal politicians whose campaigns you've contributed to
14. Your attorney
15. Your physician/dentist
16. Your minister/rabbi
17. Neighbors
18. Colleagues you know through work
19. Local school personnel you know through your children

20. Officials of professional organizations
21. Pharmacist
22. Your banker
23. Your investment counselor
24. Your accountant
25. Your hairdresser/barber

As you begin making contacts, hearing about jobs, and following up with phone calls, every call to a prospective employer should be followed up by a resume. Every one of those personal contacts also should receive a copy of your resume.

You see, most people like to help: some because it just makes them feel good, others because they realize they could very well be in your situation someday. But you must also remember that the people you've asked to help are your friends, so don't put them on the spot or make them feel uncomfortable.

CLASSIFIED ADS

While you're out there making contacts and searching for interviews, don't neglect other possibilities, however remote. For example, take a look at the classified ads. Although only about 15 percent of jobs are filled by people who answered ads, that still means that there are 15 people out of every 100 who landed a job by following up an advertisement. Answer ads with a resume and a cover letter carefully tailored to the listed requirements.

In addition to local newspapers, you should also check out major publications that serve regional or national markets: papers such as *The Wall Street Journal*, the *New York Times*, the *Chicago Tribune*, the *Los Angeles Times*, and the *Washington Post*. The *National Business Employment Weekly* (published by *The Wall Street Journal*) provides a weekly consolidation of ads appearing in the *Journal's* regional editions.

Be sure to read all the ads from A to Z, since positions may appear under a variety of headings. Cut out or copy the ad to which you're responding for future reference and be sure to tailor your cover letter to the ad.

Don't worry if your qualifications don't match up exactly with the description outlined in the ad, you might still be a strong candidate. And even if you're not quite right for that position, it could be that a person with your credentials might be perfect for a separate, unadvertised position within the company.

Still, keep in mind that you won't be the only person answering that ad. You'd be surprised at the number of people who rely solely on this method of finding a job, and so there's likely to be stiff competition for any advertised spot.

AGENCIES

Don't overlook employment agencies and executive search firms when job hunting. Here's a quick rundown on the different organizations:

Temporary Employment Service

This agency is paid by an employer to find someone to fill in temporarily for a specific length of time. "Temping" offers job seekers a terrific opportunity to try out a job before actually making a full-time commitment, and you'd be surprised at how often a temporary job can turn into a permanent position if the employer is happy with the way you settle into the organization. Temporary positions are also a good way to earn a salary while searching for a permanent position. There are even a few temporary agencies that specialize in filling consulting assignments for executive-level personnel. Temping is an especially good way for older workers to gain entree to employers; once "in the door," you can demonstrate what you can do.

Employment Agencies—Fee Structures

Using an employment agency to help you find a job can be effective, depending on the type of skills you have. Employment agencies are most useful for those positions that require specialized skills or experience in short supply on the market.

I don't advise paying a fee to anyone to help you find a job. When looking at employment agencies, consider only "fee-paid" positions. Employers utilize these agencies for hard-to-fill jobs and are willing to pay a fee to find the right person.

Especially stay away from those agencies who advertise in newspapers and promise jobs of all kinds for a fee (usually about $100). I also advise against using career counseling firms that charge thousands of dollars. Look at employment agencies as people who work *for employers;* make yourself available to them.

Employment agencies who are paid by employers have two types of fee arrangements: contingency (fee paid when placement is made) and retainer (part of fee paid when search begins, and balance paid as the search progresses).

You'll want to make sure any agencies you deal with are responsible organizations. Do they belong to the National Association for Personnel Consultants? Are they state licensed? Do they specialize in your particular area of interest?

Most employment agencies are reputable, ethical businesses, but it doesn't hurt to check. Ask your friends if they've had contact with the agency you're considering; see if you can get a reference to a good agency.

Meet the agency manager or owner. Is this the type of person with whom you want to deal? Finally, read over any document they give you very carefully—especially before you sign anything.

Executive Recruiter

These agencies are usually state-licensed to collect a retainer from a company to conduct a search for the person that the company needs. Usually, the positions the recruiter tries to fill lie in the $60,000 or higher salary range. Most companies use such a "head hunter" when the position they're trying to fill is at a very senior level, or requires very specific or special qualifications. You may send your resume to executive recruiters, along with a cover letter stating your objectives, but don't be discouraged if you don't receive an answer—your documents could be retained in the firm's files for future reference. If time and opportunity allow, a personal visit may be better than a letter.

Although the executive recruiter is an excellent source for senior-level positions, such agencies represent a very small segment of the job market. Many executive-recruiting firms are small operations themselves and may be conducting a position search for only a few jobs at a time. Therefore, I recommend that you work with several different recruiters. Browse through *The Directory of Executive Recruiters* (Kennedy Publications) for specific information.

COLD CONTACTS

You can also try the direct approach. There are two ways to handle this: Drop in without an appointment, or pick up the phone. In either case, be ready to provide your resume. If you are lucky enough to get an interview on the spot, attach your resume to the completed application and give it to the interviewer (buy a little pocket stapler and keep it with you, so you can staple your resume to the application). Or, you may be able to see the interviewer without the application if you say: "I have my resume with me."

Let's face it, there is some bias against an older job seeker, and one way of overcoming the problem is to make many, many contacts. My rule of thumb is, if you're not getting rejections, you're not making enough contacts.

But remember, the mailed resume alone—with no phone call or no "dropping in"—is ineffective. Employers are bombarded with unsolicited resumes and don't pay much attention to them. You've got to be willing to put in a little extra effort to make a strong impression.

CONDUCTING YOUR JOB SEARCH

You want to be methodical about your job search, and do the most thorough job you can. A little bit of time spent on research will make all the difference. You'll want to identify what prospective employers do, where they are located, the size of the company, key personnel. Engaging in research is also a good way to uncover even more prospective employers.

The most efficient way to do research is at your local library; you'll find it's one of the best and most underutilized sources of information for job seekers.

Start your research at the reference desk. It is the reference librarian's job to know exactly what the library has in its collection that may help you, and how to access it quickly. (For instance, the very helpful *Occupational Outlook Handbook* is often kept out of sight behind the reference desk, and you'd have to ask specifically for it.)

You'll want to check out the library's "vertical files" on a variety of careers. These special files often include briefs, literature, bibliographies, and magazine articles about the career in which you're interested. In addition, many larger libraries carry the latest federal job listings and a wide number of publications that can be a terrific source for the over-50 job hunter.

Don't overlook current periodicals, which are often more up-to-date than books. Help-wanted ads, articles on careers, job hunting, training, applying for jobs, and so on may all be included in trade journals, professional magazines, and popular periodicals, or industry directories (such as the Standard Directory of Advertising Agencies; Best's Insurance Reports; Rand McNally Banker's Directory). In addition, there are a number of valuable reference books you'll want to take a look at. These include:

Job Hunter's Sourcebook (Gale Research, Inc.)

Arranged by occupation, this includes sources of help-wanted ads, placement and job referral services, employers, employment agencies.

Million Dollar Directory (Dun & Bradstreet Corp.)

Top 50,000 companies, with descriptions and contact information. Listed alphabetically, geographically, and by industry.

State Manufacturers' Directories (Manufacturers' News, Inc.)

A state-by-state directory with businesses listed alphabetically, geographically, and by industry. These directories can be your best resource for locating employers within your community.

Encyclopedia of Associations (Gale Research, Inc.)

22,000 organizations of all types. Many industry organizations have placement services; some can provide lists of member companies.

Consultants and Consulting Organization Directory (Gale Research, Inc.)

More than 14,000 consulting firms listed by industry and location. Consulting is ideal for some over-50 job seekers.

Thomas Register of Manufacturers (Thomas Publishing Co.)

140,000 specific product manufacturers with contact information.

The Directory of Executive Recruiters (Kennedy Publications)

Recruiters arranged by retainer (employer pays fee in advance) and contingency (employer pays fees when placement is made). Also lists recruiters by industry and geographical area.

Another valuable resource at most community libraries is the bulletin board near the entrance, which often lists local government job openings and upcoming job-hunting workshops. Brochures for local colleges may also be posted here. It's also possible that your local library offers typewriters and computers available for community use—not to mention low-cost photocopiers.

Now that you have made a good start on researching job possibilities, it's time you paid attention to the way in which you're managing your job search. You've been managing your job—and maybe managing machines, people, and inventory—for more than 20 years. By now, it should come naturally to you.

- *Prepare a daily and weekly schedule.* Decide how many contacts to make, how many resumes to send, how many follow-up calls to make.
- *Get face-to-face meetings.* Call them "interviews" if you want to, but it pays to talk to companies who are not now seeking to fill an opening. The over-50 job seeker has so much to offer. The employer may actually create a job for you.
- *Get your resume to a broad network of readers.* The more people who see your resume, the more people will know of your skills and experience. Increase your chances.
- *Don't be discouraged, don't be distracted.* You'll be making lots of contacts. Many of them will say "no." You have to be willing to get some "no's" in order to achieve a "yes". Job hunting isn't the most pleasant work in the world, and it's easy to get distracted by jobs around the house, by soap operas, or by a convenient tee time. Give those distractions second priority. Continue to have some fun, *but* your job search comes first!
- *Avoid personnel departments.* (What, Sam Ray? After 35 years as a personnel manager?) I can tell you honestly that personnel managers recruit, but it's the person in charge of the department (the engineering manager, the store manager, the office manager, the vice president) who makes the hiring decisions. Do some research to find out who's in charge, and get your resume to that person.
- *Turn over every stone.* Been turned down by one company? Contact them again—it shows you're determined and interested in that particular company. Perhaps the person they hired for the last job didn't work out, or maybe there are new positions opening up. You won't know if you don't check it out, and since you've already been interviewed once, they know a little bit about you.

11

Managing the Interview

By this point, you've created a resume that succinctly sums up your skills, experience, and abilities. If you've followed the techniques I've outlined, your resume will probably have opened some doors for you by now, and you've gotten some interviews lined up.

All my earlier recommendations—taking an inventory of your skills, targeting your objectives, researching companies, identifying contacts—have helped prepare you for the moment you walk in for your interview.

It's important to understand that while a resume is an objective, straightforward catalog of what you do best, an interview is just the opposite: It's a completely subjective assessment of *how you act* more than what you say. It's really an audition, and how well you play your part—your personality, confidence, mannerisms, gestures, looks, poise, fluency—will determine whether you get the role.

The over-50 job seeker can be particularly vulnerable in interviews by having either too much or too little confidence. Many older workers are so experienced they decide they don't need to prepare for an interview. Others approach an interview with a negative attitude: "It's been 14 years since my last interview. *I'm not up to this!*"

Both attitudes are unhelpful. The fact is, *anyone* can carry off an interview. You simply need to understand the process and prepare yourself. Know the rules, and play to win.

TYPES OF INTERVIEWS

There are four types of interviews, and you'll want to be prepared for them all.

Tell-Me-about-Yourself Interview

You'll recognize this interview style immediately because you'll probably hear the question within the first few minutes. Beware: This isn't as friendly and disarming as it may seem; it's easy to go wrong here. Just in case this is the kind of interview you end up with, prepare ahead of time by rehearsing a two-minute talk about your background: where you grew up, your education and early work life, and your most relevant and recent experience. Then ask the interviewer what else he or she would like to know. Encourage the interviewer to ask specific questions, and then ask the questions you want to know. By practicing beforehand, you won't be thrown by this off-the-cuff question, and you won't end up floundering around for something to say.

Listen-to-My-Story Interview

You'll know you've embarked on this kind of interview when you realize the interviewer hasn't stopped talking since you walked in. Don't let him or her natter on endlessly; if you don't manage to break in, you won't make any kind of dynamic impression. Interrupt politely: "Excuse me. I'm interested in that project, because you know—it sounds similar to one I headed up on my last job. Let me tell you a little about what I did . . ."

Stress Interview

The most difficult of the four types and fortunately uncommon, the stress interview may involve one or more interviewers firing questions to see how you handle pressure. Whatever you do, don't lose control. If you find yourself slipping under, *stop*. Take a deep breath and, keeping a smile on your face, say something like: "Wait a minute. Let's take one question at a time. I'm here to give you an interview, not a confession. Now, Ms. Taylor, you were asking me about my background in . . ."

Regular Interview

No unusual stress, nasty questions, or evasive interview techniques here. This is just a straightforward, to-the-point information session aimed at learning a little about you to enable the company to fill a position with the best candidate.

RAY'S RULES FOR SUCCESSFUL INTERVIEWS

First: Be Prepared!

When scheduling interviews, remember that you'll want to be fresh and in tip-top form, so don't squeeze four or five meetings into one marathon day. Try to schedule just two interviews a day, one in the morning (9:00 or 10:00) and the second in early afternoon. This way, you'll be alert for the first session, and then have a break to reread your research, eat, relax, and prepare for the second round at 2 P.M. It's a good idea to do this three or four days a week, with one day left for follow-up calls, letters, and more scheduling.

You will also avoid having to go out on a luncheon interview, which offers too many opportunities for something to go wrong. (Spilling food or using the wrong fork doesn't have anything to do with your work, but it will cause a bad impression nonetheless). You'll probably end up sitting down to a working meal at some point in the interview process, but it's best to put that off as long as possible.

Do Your Homework

Only a neophyte would go into a meeting as important as a job interview without doing some research. Once you've scheduled an interview, you'll want to start an intensive study of the company. Review your research notes, call contacts to see what they know, and even make anonymous calls to the company to gather more data. What are you looking for? Obviously, you'll want to know the basics (product line, market share). This way, you can speak intelligently during the interview and show you wanted the job enough to do your homework. But it's even more important for you to get an understanding of the *company culture:* its values and the kinds of employees it hires. That's what the interviewer will be looking for during that first interview: Will you fit in with the team, or be a mismatch?

Dress for the Part

It's a good idea to reread Chapter 2, where I discussed proper interview attire. You want to present a vital, energetic, positive, and

Figure 11.1 Checklist-Your Appearance

- ☐ Get a good night's sleep. Use eyedrops to keep your eyes clear.
- ☐ Evaluate your weight. If your diet isn't working and you're a few pounds overweight, camouflage with well-tailored clothes. They cost more, but they're worth it.
- ☐ Take a good look at your hair. If it's gray and distinguished, fine. If it's gray and washed out, cover it up with a temporary wash-in hair color. And if you're balding, accept it. Attempting to cover it with an awkward hair style or a bad toupee creates a bad impression.
- ☐ Keep makeup, jewelry, and hairstyle discreet and understated. You want to focus attention on you and what you're saying, not on accessories that scream "bad taste."

well-groomed appearance, *especially* if you're over 50. You can bet that if you walk in the door "looking old" (huffing and puffing, overweight, out of shape) the interviewer will silently note those facts. It's unfair, but it's a fact of life, and if you want a job you must play by the rules.

Try to dress so that you're as relaxed as possible about your appearance rather than self-conscious or uncomfortable. An otherwise minor irritation, such as a pinching waistband, too-tight shoes, or bunching jacket, can interrupt your concentration and interfere with your presentation. Figure 11.1 is a checklist to ensure your best appearance:

Attaché Tips

You'll want to make sure to include everything in your attaché case that an interviewer could possibly ask to see. (And of course, that case should be as presentable as the rest of your wardrobe: high-quality material in good repair, preferably matching shoes and other accessories).

Here's what to include:

- Resumes: Six copies on quality bond stationery
- References: Names, addresses, contact numbers of three or four supervisors and other professional contacts
- Letters of reference: Optional, but a few recent reference letters addressed to "To Whom It May Concern" are acceptable. (The best letters, however, are written directly to your prospective employer by an influential reference after you've been interviewed.)

- Work samples
- Special licenses
- Names of all former supervisors
- Legal pad and two good ballpoint pens

Punctuality Plus!

After you've spent all this time preparing for the interview, the last thing you want to do is arrive late. Allow plenty of time for snarled traffic and parking problems. If you find you're too early, don't appear at the receptionist's desk any earlier than five minutes before your appointment. Instead, take a stroll around the grounds or office building or step into the restroom to check your appearance. If you're running late, call; if you're running more than 15 minutes late, call and reschedule rather than risk an irritated interviewer.

The First Five Minutes

Psychologists tell us that indelible impressions are made during the first five minutes of an introduction, so you'll want to work on the way you begin your interview. Practice your initial greeting in front of a mirror, or get a friend to role-play with you. It's a great idea to get a third person to videotape the entire encounter, practicing until you get it right. By practicing, you'll be able to reduce tension and therefore lessen the chances of saying something wrong. If you seem at all uncomfortable or awkward, you'll have made a negative impression.

Both men and women should shake hands with a firm grip, look the interviewer in the eye, and say something like: "Good morning, Mr. Jones. I'm Kathryn Ellerby. I'm pleased to meet you." Speak slowly and enunciate, especially if your name is at all difficult to pronounce. Smile with your eyes and your mouth, and don't worry about witty repartee right in the beginning. Allow yourself time to settle in and get the interviewer's attention.

Look around for something to comment on (an award, a wall hanging), but avoid personal items such as photographs. Keep your comment brief; you want to establish rapport, but you're really there to talk about *you.*

If this sounds like a personality contest, you're absolutely right. That's exactly what it is, and winning it means getting a job. No matter how experienced a businessperson you are, even the most seasoned pro can benefit from a bit of practice.

Here's what your prospective employer will be trying to decide during the interview:

- Are you neat and businesslike?
- Do you answer tough questions with sincerity and conviction?
- Do you have necessary information at your fingertips? (names, dates, references)
- Are you on time?
- Are you convincing?
- Will you fit in with co-workers?
- Do you seem easy to get along with, pleasant and personable?

Don't Smoke or Chew Gum

Even if the interviewer is knee-deep in ashtrays and puffing on a cigar, don't light up yourself, even if specifically invited to do so. It's not a social visit, it's a business presentation and you don't want the distractions involved in smoking. Besides, these days smoking in the workplace is getting a bad name. Especially in those over 50, a smoking habit could make an interviewer suspect you might be in poor health. Don't have smoking materials in evidence and don't smoke in any confined area before the interview—the smoke will cling to your clothes.

Beware the Desk

The interviewer's desk is an authority symbol that gets in the way of establishing rapport. When you enter the office, try to prevent the interviewer from disappearing behind this barrier. If there is a couch or two comfortable chairs side by side in the office, try to sit there with the interviewer. You might try luring him or her away from the desk by asking questions about the facility visible from the window. Could you tour the area where you would be working, or meet the person who would be your immediate supervisor?

If the interviewer doesn't seem inclined to venture from behind the desk, try walking around the desk to show samples of your work but respect the interviewer's "comfort zone" and don't move in too close.

Watch Your Attitude

Speak clearly, directly, and politely, and always look the interviewer in the eye. Interviewers are trained to pick up nonverbal cues, and failure to make eye contact can be interpreted as dishonesty. Sit up, lean forward and try to be enthusiastic. Watch your posture: You want to appear confident, not bored or nervous. Be positive and never criticize your former job, boss, or colleagues.

If you had a rough time in your last job, don't describe it as a "personality clash"—refer instead to "policy differences." If you seem cranky or bitter about your last job, it's going to reflect badly on you, not them.

If in Doubt, Ask

If you don't understand something the interviewer has asked you, don't be afraid to ask the interviewer to rephrase it. And once you've answered, don't be afraid to ask whether your answer was satisfactory.

Don't Beg

Too many job seekers end up begging for a position. Even if you feel as if you would take *any* job somebody tosses your way, don't let this attitude show during the interview. Before you accept the position, try to have all the facts so you can make a good choice; this might be the last career decision you'll have to make.

You want to be discreet, so don't immediately ask about salary, benefits, and vacation policy. Don't pepper the interviewer with questions up front, but work in some intelligent questions during the course of the session. This shows you're interested and gives the interviewer a break.

Here's an idea of the kinds of things you could ask:

- To whom would I report?
- What positions would report to me?
- What is the job description?
- What is the size of the department?
- What is the department structure?
- Where would I be working?
- Why is this job vacant?

- How long has it been open?
- How many have applied for it?
- Are any current employees applying for it?
- What is the turnover rate for this job?
- How much travel is involved?
- Will I be trained? What kind?
- What about performance reviews?
- How are raises and promotions decided?
- What kind of growth and opportunity can I expect?
- What is the corporate philosophy?
- Are there plans for expansion, relocation, or closings?
- Are any of the workers unionized?
- Are sales up, down, or the same?
- When did the present owner buy the company?
- Is there a takeover or merger pending? Is so, how will this affect present management?
- When will you make your decision?

Of course, if you've researched the company, you'll already know the answers to many of these questions. Ask the ones that seem appropriate, but don't interrogate and don't insist on answers if the interviewer seems put off by your line of questioning.

HANDLING INTERVIEW QUESTIONS

The hardest questions you're going to have to deal with are those that focus on your age. There are many ways to phrase an age question, but they all boil down to the same basic issue: "I think you may be too old for this job." Your job is to convince the interviewer that this basic idea is wrong. You can't evade the issue, so the best idea is to prepare for these questions head-on.

My favorite disguised version of the age question is this one: *"You seem overqualified for this position."*

Here's what I'd say if somebody said this to me:

> I'm glad you perceived my strong qualifications right away. I'm probably more qualified than some other candidates, but I don't believe I'm overqualified. I'll be able to take over the position with little or no down time, and I'll be able to use my experience to bring

more to the position than you've come to expect. That way, I'm challenged and the company gets a better return on its investment.

The response to this is sometimes, *"But perhaps a younger person . . ."* You can face this issue with these comments:

> I sense you're concerned about my age. I'm in excellent health and in the prime of my career. I'm not ready to put my knowledge and skills on the shelf for many years. There's a lot I still want to accomplish, and if this company hires me, it will be getting the benefit of my experience and knowledge. That's an asset, not a liability.

An interviewer is going to want to know a variety of other things about you in several areas. Here are many of the typical questions you should prepare for:

- What qualities and skills do you think the successful candidate should have for this job?
- Why do you want to work for us?
- What can you bring to this job?
- Why should we hire you?
- How do you feel about leaving your last company?
- Why haven't you found a job yet?
- Does anything about the job bother you?
- What interests you about us?
- Tell me about the positions you've held.
- What things in your work experience have you enjoyed the most and the least?
- What kinds of outside activities are you interested in?
- Tell me about problems you've solved on the job.
- What new skills have you learned on the job?
- Tell me about your biggest success at work.
- Are you willing to relocate or travel?
- Can you work long hours?
- Why did you leave your last job?
- How did you get your last job?
- What are your plans for the rest of your career? Where would you like to be in the next five years?
- What are your current earnings? What do you expect to earn with us?

- What is your definition of success?
- Are you satisfied with your career progress to date?
- Why haven't you advanced farther in your career?
- What type of work are you looking for?
- How do you plan and organize your work?
- Describe the best manager you ever had.
- What would your former supervisor say about your strengths and weaknesses?
- How would your subordinates describe you as a manager?
- Give five adjectives to describe yourself.
- What is the definition of a good manager?
- Have you ever fired anyone? Why? How did you go about it?
- If you had your boss's job, what would you have done differently?
- Describe a subordinate of whom you are most proud. What did you have to do with that success?

THE BOTTOM LINE: SALARY

There's no hard and fast rule about who should bring up the subject of compensation. With your experience, however, you probably have some idea of the salary range for the position for which you're interviewing. If not, there are a number of ways to ascertain a salary range before the interview:

- Check professional journals, which often publish salary surveys.
- Read the National Business Employment Weekly.
- Investigate professional organization salary surveys.
- Ask others in your network.
- Ask people at the employment agency.

Go into the interview with a firm idea of your minimum requirements. Will you have to make a move? Will you be losing benefits? If you must relocate, what about different housing costs and standard of living? Selling a house in Nebraska and moving to San Francisco means a significant change in economic situation. Will the new job cover the difference in costs?

If the interviewer doesn't mention money, you can bring up the subject. Ask: "What is the salary range for this position?" (Keeping in mind, of course, that a "salary range" is abandoned every day when the right candidate comes along).

If an interviewer asks you: "What kind of a salary are you looking for?" You don't have to feel as if you're on the spot. Instead, respond: "What is the range for this position?" or "What is the salary range for an experienced person?"

At the same time, keep in mind that money isn't everything, and a more modest job may be better than no job at all. With your years of experience, your salary level may be so high that it is beyond the reach of a prospective employer. In this case, try to negotiate for other kinds of income: a 90-day salary review, company car, bonus plan. In my own experience and in advising Transition Team candidates, I've found that this can work.

AFTER THE INTERVIEW

Jot down key points: full name and title of the interviewer(s), what happens next, points you want to make in your thank-you letter, when the company has promised to contact you. If you haven't heard from the contact person within that time, call.

Prepare your inquiry in advance (repeat the main points from your thank-you letter, and ask for the job status). The purpose of this call is to remind the employer of your interest as well as to find out where you stand.

After every interview, get on the phone with other prospective employers and try to get more interviews. Keep going out on interviews until you get that new job.

12

Go "High Tech"

While computers are a real boon in many ways, when it comes to writing a resume, it's really better to create your own (using the information in this book as a guide) than to let a computer create one for you. However, if you really aren't satisfied with what you've come up with on your own, there's no harm in seeing if a software package can help you out.

Like many in my generation, you may have "missed" the computer revolution, but there's no need to be afraid of these machines. And in fact, the more familiar you are with computers, the more you will help yourself and make yourself seem a more up-to-date candidate.

When it comes to looking for a job, "high tech" means having access to FAX facilities, being able to get information ("information retrieval") and preparing a resume with a software package.

FAX MACHINES

There's no doubt about it—almost everybody in the business world today who wants to get information out fast has access to a FAX machine. You can only help your situation if you do, too. Which of the following scenarios would be more impressive if *you* were doing the hiring?

> Employer or agency (on phone) asks for your resume ASAP. You mail it and it's received two days later in the daily stack of mail. Or you send it overnight mail and it arrives the next day, costing you $9 or more . . .
>
> Or

Employer or agency (on phone) asks for your resume ASAP. You ask for the FAX number. You hop in the car, go to the nearest FAX machine (nearby copy shop, quick-print business, hotel lobby, etc.). You fax your resume, and it's received instantly. Your cost: less than $5.

INFORMATION RETRIEVAL

FAX machines aren't the only technical assistance you can call on for help in finding a job. Computers can be essential, and these days, they aren't used just for number crunching. If you want to compete effectively in today's market, you must be as creative and aggressive about finding a job as possible. Computers are a helpful tool.

The key to finding a good job is information: Who is hiring, what are the company's problems, what are the company's competitors doing, what might the company be doing in the next five years (both administratively and productwise)?

It would take hours of tedious research in a library to find the answers to all these questions, but this kind of fact-finding can be done in about an hour using computerized information retrieval. With a computer, you can cover the full range of business publications (there are more than 100,000) from wide-circulation magazines to little-known technical and trade journals.

A computer retrieval service can provide you with a tailored list of prospective employers. You develop your criteria by size (annual sales), location (only the states in which you are interested), and industry products manufactured or services provided; and in minutes you can receive a list of appropriate companies, with names and addresses of key executives.

What other kinds of information can you uncover?

- A brief synopsis of an article published as recently as two days earlier
- Details about who wants to buy out a company—and why
- Upcoming major management changes
- A peek into a company's advertising campaign
- Product market shares and potential new developments
- The scoop on major competitors (planned production programs, strategies, etc.)
- Most recent press releases
- Information about key executives (including salary)

- Detailed financial information
- Executive biographies
- Speeches
- Data concerning cities (for possible relocation)

Obviously, once you have this information, it's up to you to use it to your advantage. But with the right inside information, you can anticipate questions and be prepared to respond with thoughtful answers that demonstrate your grasp of the company's problems. With the most up-to-date financial information at hand, you're also in a much better position to negotiate.

How do you go about getting access to information with a computer?

1. Your personal computer at home. You need a PC and a modem. A modem gets you direct access, via telephone lines, to the information. Your neighborhood computer store can supply the modem and the software. The next step is to call a number of data-retrieval services to shop around and find out exactly what kind of information will be provided and how much it will cost. The following are some examples of data retrieval services:

   Compuserve
   Columbus, OH
   800-848-8199

   Prodigy Services Company
   White Plains, NY
   800-776-0840

   News Retrieval (Dow Jones)
   Princeton, NJ
   800-522-3567

   Dialog
   Palo Alto, CA
   800-334-2564

   NEXIS (Mead Corporation)
   Dayton, OH
   800-227-9597

2. If you don't have a computer at home. In this situation there are two approaches:

   - Your local library and/or college library. Do you, the 50+ job seeker, have fears about using a computer terminal?

The librarian will help you conquer those fears. The information retrieval procedure can be learned in just minutes. Many libraries have computer terminals with access to such databases as:

Standard & Poor's: Information about companies

Businesswire: Newspaper articles

INFOTRAC: Magazine articles

Electronic Yellow Pages: Information about construction, financial services, manufacturing, retailing, wholesaling, and services companies.

➤ Buy the service. Some firms will provide the information in written form. For example, at NEXIS Express (800-843-6476) a NEXIS expert will conduct your information search for you. NEXIS will then mail the information to you; if you're in a hurry, it can be faxed or sent overnight express mail.

COMPUTERIZED RESUME PREPARATION

If you think you'd like to try to put together a resume with the help of a software package designed to walk you through the process, a wide variety are on the market for about $50 or less.

In general, computer resume-creating packages help you design an impressive-looking resume and write cover letters to go along with it. The packages usually offer a variety of resume templates (with job-specific categories) that allow you to plug your specific job information into their format. But because computers are inherently powerful, their designers didn't stop there; most software packages also help you track job leads and interview appointments.

The problem with these packages is that they *are* generated by a computer, hence, there is very little flexibility. Some have awkward formatting or spell-checking features. One package in particular limits the field for inserting the company name to 31 characters, including spaces. If your company's name is longer, you're out of luck. That's where the "inflexibility" part can really be a problem.

Still, they may come in handy if you really feel stuck on your own. If you think you'd be interested in checking out a software package for resume design, look in the Yellow Pages of your phone book under "Computers—Software and Services" or contact:

Individual Software, Inc.
San Carlos, CA 94070-2704
(800) 331-3313
(800) 874-2042, in California
Program: *Resumemaker*

Spinmaker Software
201 Broadway
Cambridge, MA 02139
(617) 494-1200
Programs: *Resume Kit, PFS Resume, Job Search Pro, Better Working Resume Kit,* and *Easy Working Resume Creator*

Boutware Software, Inc.
Agoura Hills, CA
(818) 706-3887
Program: *Professional Resume Writer*

Gregg, Division of McGraw-Hill
New York, NY
(212) 512-6665
Program: *Business Resume Preparation*

Sourceview Software, International
Concord, CA
(415) 685-3635
Program: *Pro Resume*

Cambridge Career Products
Charleston, WV
(800) 418-4227
Program: *Resu-Riter*

PART IV

SAMPLE RESUMES

13

RESUME CHECKLIST

The following checklist and the resume samples will help you develop your own resume.

- ☐ One page, two pages maximum
- ☐ Clean type
- ☐ Good balance
- ☐ Adequate white space
- ☐ One-inch margins
- ☐ Chronological format, most recent position first
- ☐ Strongly worded career summary at top
- ☐ Reads easily
- ☐ Crisp, clear sentences
- ☐ No long paragraphs
- ☐ Bulleted lists
- ☐ Important points highlighted in boldface (not too many)
- ☐ Dates, titles, and employers discernible at a glance
- ☐ No spelling, grammar, or punctuation errors
- ☐ Uses action words
- ☐ Accentuates positive
- ☐ Stresses results
- ☐ Stresses accomplishments
- ☐ Uses dollars, percentages, numbers
- ☐ Is a persuasive "preview of coming attraction"
- ☐ No salary data
- ☐ No "reasons for leaving"
- ☐ No dates or personal data that give away age
- ☐ No misstatements or exaggerations

RESUME SAMPLES

The samples on pages 109–208 represent well-written resumes selected by me and Gail Ryder, my Transition Team colleague, from our files of hundreds of resumes. We've done a little editing, mostly to camouflage names and addresses.

Some of the samples don't exactly reflect all the rules and suggestions I've included in this book. Resume writing is more an art than a science. But all the resumes were written by job hunters who were 50ish, and over. I think you'll agree that none of them contain age-specific information that would turn off prospective employers.

Of course, no sample will exactly fit your background—after all, it's taken you 20 years or more to develop your unique experience profile. But the samples can get you started, act as a guide, and can demonstrate the ideas I've described in earlier chapters.

BARNEY B. HELMS
367 Frost Lane
Franklin, Tennessee 37064
(615) 791-4152

Vice President Finance and Administration with extensive experience in accounting, finance and computer systems.

Demonstrated record of accomplishments with both a Fortune 500 manufacturing company and a large public accounting firm. Financial expertise includes:

- Strategic Planning
- Management Accounting
- Capital Budgeting
- Computer Modeling
- Financial Reporting
- Operations Analysis
- Acquisitions
- Computer Controls

PROFESSIONAL EXPERIENCE:

GENERAL CEMENT CO. 1979-Present

Vice President Finance and Administration - Mississippi Valley Operation

- Direct finance, accounting and administrative activities for operations with annual sales of $150 million.
- Responsible for all accounting functions through review and approval of financial information before presentation to the coporate office.
- Directly responsible for the supervision of 18 person staff.
- Responsible for decentralization of mainframe computer system to IBM AS400 system.
  *Developed Profit Center's business strategy for expansion.

Senior Director Asset Management - Dallas Processing Center

- Managed corporation's major assets including cash transfers, accounts receivable credit and collection ($33 million), inventories ($68 million) and capital assets and budgets($600 million).
  *In-depth analysis and assessment of accounts receivable resulted in the collection of an additional $2 million.

Director of Auditing - Corporate Office

- Managed and coordinated all audit activity for domestic locations, international subsidiaries and corporate office.
  *Developed and implemented computer auditing procedures.

DURSTINE ACCOUNTING ASSOCIATES
1977-1978 Director of Accounting and Auditing - Brunswick, Georgia

BDO SEIDMAN
1967-1977 Manager, National Office - Cleveland, Ohio

Manager, Denver Office

EDUCATION: BSBA in Accounting, University of Denver; Certified Public Accountant; IBM Systems Science I Institute

PROFESSIONAL AFFILIATIONS: American Institute of Certified Public Accountants. Past member of AICPA Taskforces: On-Line Systems, Audit Impact of Complex Computer Systems

PERSONAL: Married, excellent health, willing to relocate

ANDREW P. GERALDO

3748 Apple Cross Boulevard
Louisville, Kentucky 38982
(478) 938-3095

Over 20 years experience in the Plant Financial Controller Department for a major multi-national automobile manufacturer.

EMPLOYMENT HISTORY:

Acme Corp., 1001 North Highway Drive, Louisville, KY 38982
(Available due to plant closing.)

Financial Analyst

Manage the forecasting of capital spending and preparation of the Capital Investment Management System (CIMS) reports. Also analyze project requisitions for proper line item reporting and control, and write briefs with information on return of investment, depreciation recovery, income tax credits and labor saving summaries. Additional duties are coordinating the lease car, car product evaluation and company car control programs.

* Implemented 32 on-line computer project control systems, which exceeded the department goal.
* Member of a cost savings team that saved over $1.5 million dollars in the construction of the Industrial Waste Treatment Plant. Received a letter of commendation from the Corporate President, James P. Robertson.
* Saved over $150,000 in the Paint Department by changing the piping which eliminated the loss of paint.
* Reduced the plant-wide compressed air losses, savings over $160,000. Received Management award.

EDUCATION:

B.S.B.A., Transylvania University, Lexington, KY - Accounting Major

Acme Institute
Training in Management, Motivation, Computerized Data Recovery Systems and Corporate Courses on Budget Control and Accounting

MILITARY SERVICE:

Kentucky Army National Guard - Honorable Discharge

HOBBIES AND INTERESTS:

Golf, Bowling and Children's Sport Activities

VOLUNTEER WORK:

Subdivision Trustee

OLGA C. HOPPER

3852 Peninsula Ct.
Sarah, Missippi 38664
(601) 750-6137

PROFESSIONAL EXPERIENCE

Auto Club of Mississippi, Jackson, MS, 1981 - Present
Financial Reporting Supervisor

Managed staff of senior accountants, accountants and staff accountants responsible for asset reporting for 12 Auto Club of Mississippi companies.
Reported investment portfolio of more than $1 billion.
Reported, maintained and monitored cash book balances for various multi-million dollar inter- company bank accounts (annual premium receipts of $500 million).
Reported and budgeted fixed assets and analyzed property related expense variances.
Developed inter-company property charges for all facilities and cost centers.
Represented Finance department on corporate task forces that created new systems and maintained existing ones.
Analyzed and budgeted staffing needs.
Recruited and hired candidates, conducted performance reviews.

Financial Statements Supervisor

Managed staff of senior accountants and accountants.
Prepared financial statements for 12 Auto Club of Mississippi companies (4 of which are insurance companies) using Dynaplan spreadsheet package which is similar to Lotus 1-2-3.
Other financial reports included change in balance sheet analysis, cash flow and expense variance analysis.
Prepared insurance annual statement.

Disbursements Supervisor
Supervised accounts payable staff of 13.
Managed cash disbursements system, coordinated system changes and updates, maintained internal controls.
Chairperson of Uniform Accounting Committee.
Contact person assigned to resolve upper management concerns and inquiries related to departmental expense charges.

Kroger Dairy (presently Mississippi Dairy), Livonia, MI 1977 - 1981
Assistant Plant Controller

Supervised accounting staff through all phases of computerized accounting system through trial balance and issuance of financial statements.
These systems included payroll, raw material and finished product costs, material utilization, accounts receivable and payable, property accounting and productivity analysis.
Coordinated and prepared comprehensive plant budget.
Converted company reporting to detailed reporting by product.

Accounting/Accounts Payable Clerk

Handled accounts receivables and balanced payroll accounts.

Previous positions included: Audit Department Supervisor, Auditor, and Customer Service

EDUCATION

B.S., Business Administration	University of Mississippi
Major: Accounting	Oxford, MS

JESSICA L. RAYMONDS

31020 Franklin Street
Ann Arbor, Michigan 48766
(313) 987-4950

EMPLOYMENT HISTORY

ACCOUNTING CLERK

(Available due to corporate reorganization.)
Alco Department Stores, Inc., 4500 W. Jefferson Avenue, Freeport, MI 48768
1978 - Present

Responsible for handling Accounts Payable functions for a chain of 14 retail department stores throughout Michigan. Prepare Accounts Payable invoices for payment, check prices and extensions for possible errors, prepare purchase debit memo in case of error.

Key information from distribution ticket into computer to distribute cost and generate check to vendor.

Other duties included:

* Operating Switchboard * Sorting Incoming Mail

* Posting Outgoing Mail * Writing Inventory Tags

* Typing & Filing

* Outstanding work, attendance and punctuality records.

* Enjoy a busy work environment.

EDUCATION

Pontiac Business Institute Pontiac, MI	Accounting and Secretarial
Western High School Detroit, MI	Graduated Commercial Courses

HOBBIES AND INTERESTS

Ceramics * Bowling * Dancing * Golf * Swimming

VOLUNTEER WORK

American Cancer Society * United Way * Girl Scouts of America

References Available Upon Request

WARREN PIERCE

4535 Lake Drive
Bay City, Michigan 48000
(313) 321-1234

Certified Public Accountant with extensive internal audit, financial management and public accounting experience. MBA and BS degrees with honors.

PROFESSIONAL EXPERIENCE:

1980 - Present
Ritetrack Corp.
Bay City, MI

GENERAL AUDITOR - Corporate Headquarters

* Corporate-wide internal audit responsibilities for Ritetrack Corp.: combined activities generate annual revenues in excess of $3.0 billion.
* Managed up to 12 Managers.
* Introduced operational auditing, EDP audit techniques (Culprit, Focus, Auditape and Control Plan), and the coordination of audits with external auditors.
* Instrumental in obtaining management acceptance of a computer control system. Improved controls resulted in $1.5 million annual savings.
* Coordinated four successful lawsuits with company management, outside legal counsel, and government law enforcement agencies; one lawsuit resulted in a Federal jury award of $500,000.
* Instrumental in the acquisition of three profitable businesses and disposal of five unprofitable operations.

MANAGER, FINANCIAL PLANNING - Corporate Headquarters

* Major role in acquisitions and divestitures strategy including the relocation of Company Headquarters and R & D Department. Sold two manufacturing units with fixed cost savings of $30 million annually.

1975-1980
Peat, Moss & Co.
Public Accountants
Grand Rapids, MI

SUPERVISING SENIOR AUDITOR

* Employed upon graduation from college and rapidly promoted three levels to Supervising Senior.
* Supervised up to five professional auditors.
* Handled simultaneous audits in varied industries, including manufacturing, wholesale, retail, financial, transportation, non-profit, trusts and advertising.
* One of selected auditors entrusted with initial (first-year) audit engagements.

PROFESSIONAL AFFILIATIONS:

American Institute of Certified Public Accountants

Michigan Association of Certified Public Accountants

Institute of Internal Auditors

EDUCATION:

M.B.A. - University of Delaware

B.S. - Accounting (Cum Laude) - University of Georgia

PERSONAL:

Married, three children. Excellent health.
Willing to relocate.

BERTHA A. COWBELL

87440 Hottentot Road
Sterling, Michigan 45078
(313) 839-6676

EDUCATION

1986 - 1991 Bachelor of Science, Business Administration
Managerial and Cost Accounting
Eastern Michigan University
Ypsilanti, Michigan

1975 - 1977 Accounting
Completed 66 credit hours toward degree
Macomb County Community College
Mt. Clemens, Michigan

EMPLOYMENT HISTORY

MANAGER, ACCOUNTING DEPARTMENT/ASSISTANT TREASURER
Automo U.S., Inc./Automo (Canada) Ltd., Novi, Michigan
1982 - 1991

Responsibilities:

- Supervise Accounting staff through all phases of computerized accounting system to trial balance and issuance of financial statements.
- Issuance of profit/cost center statements and analysis of departmental variances to budget.
- Segment reporting/Subsidiaries consolidation.
- Coordination and preparation of annual operating and capital budgets.
- Establish departmental policy and procedures.

OFFICE MANAGER/FULL CHARGE BOOKKEEPER
Marcova Tool & Die, Inc./Marcova Industries, Inc., Rosemont, Michigan
1979 - 1982

Responsibilities:

- Supervised Accounting and office staff in all day-to-day activities.
- Coordinated all phases of financial recording through monthly financial statements.
- Initiated complete job cost system.
- Purchased all raw materials for production, $330,000 annually.
- Issued financial reports to substantiate line of credit on a weekly basis.
- Established office policies and procedures.
- Consulted about company decisions and future planning.

FULL CHARGE BOOKKEEPER
T.S.Z. Restaurants, Inc., Detroit, Michigan
1977 - 1979

Responsibilities:

- Set up bookkeeping system for new business venture
- Hired and managed personnel.
- Purchased equipment.
- Established policies and procedures.
- Consulted about company decisions and future plans.

ASSISTANT TO OFFICE MANAGER/ACCOUNTS RECEIVABLE
Tom Graham Associates, P.C.
1975 - 1976

Responsibilities:

- Invoiced customers - included hospitals, nursing homes, insurance companies, private parties, medicare intermediary and medicaid.
- Recorded and deposited cash receipts.
- Posted and reconciled accounts receivable.
- Handled delinquent accounts collections.
- Petty Cash disbursements.
- Completed worksheets for yearly cost report for medicare returns.
- Assisted with accounts payable, payroll, personnel, purchasing, statistical reports and contracts.

SKILLS

Experienced with IBM 34, RPG programming and specialized accounting programs; IBM, EPSON, APPLE personal computers; software Lotus, Word-Perfect, Peachcalc, Valdocs, Formstool, and others.

MEMBERSHIPS

Notary Public
American Society of Cost Accountants

INTERESTS

Camping, Skiing, Painting, Gardening.

BRENDA C. CHIMPOURAS

2839 West Allen
Little Rock, Arkansas 72204
(501) 356-3049

Twenty year accounting career in a manufacturing setting, progressing from General Accountant through Accounting Manager. Skilled in the following functions: Consolidation of Financial Statements, Accounts Payable, State & Federal Payroll reports, Inventory Audits, Cost Analysis and Supervision.

PROFESSIONAL EXPERIENCE

Ajax Forging and Casting Corporation
Route 52, Howell, Arkansas 72203
(due to plant closing)

1977 - Present

Manager of Accounting
Reported directly to Assistant Controller and Controller. Supervise Accounts Payable, Payroll department and staff of 7 General Accountants.

- As a member of evaluation team, reviewed payroll system and made final recommendations.
- Implemented HR 2000 Payroll System with IBM System 38.
- Developed checking and balancing system for Accounts Payable which increased the accuracy of financial statements.
- Designed a costing system to compare actual to standard costs for production, which resulted in increased worker productivity (provided accurate basis for incentive pay.)
- Founded independent supplier corporation to provide company with a major component necessary for manufacturing. (Assisted with conversion to in-house operation, three years later.)
- Customized IBM System 38 financial package to produce income statements and balance sheets.
- Traveled extensively on corporate audits.

B & B, Inc.
426 East Shelby Drive, Holly Springs, Missouri 97387

1973 - 1977

General Accountant
Responsible for preparation of monthly journal entries, general account reconciliation, bank statement reconciliations; Federal, State, and City tax returns; cash and working capital analysis; auditing, inventories and balance sheet and income statement consolidation.

Additional positions held: Payroll Clerk, Accounts Payable Clerk, Accounts Receivable Manager

EDUCATION

Delta State University
St. Louis, Missouri

Bachelor of Business Administration
Major: Accounting

References Available Upon Request

LEE S. VALENTINE

9172 Neff Road
Lapeer, Michigan 48320
(313) 284-7623

ADMINISTRATIVE EXPERIENCE

Performed a variety of administrative functions including:

Operated 10-key and Dictaphone.
Input data, generated reports on computer and type 60 words per minute.
Controlled telephone communications and filed many diverse records.
Supervised and trained staff.

* Reorganized filing systems increasing overall efficiency.

* Able to both communicate well with others and complete tasks in a timely manner.

BOOKKEEPING EXPERIENCE

Possess broad range of skills including:

Prepared both computerized and manual financial statements.
Prepared profit and loss statements, balance sheets, and bank deposits.
Posted general ledger accounts and set up general ledgers on computer.
Entered financial data on mainframe computer systems.
Monitored accounts receivable and paid accounts payable.
Composed correspondence to follow up on delinquent accounts.
Computed payroll and state sales tax.
Gathered financial data from various sources in the company and monitored the accuracy of this information.
Reconciled bank statements.

* Commended for ability to pay close attention to details and accuracy.

* Highly conscientious, stable, dependable.

WORK HISTORY

Over 20 years of experience, including the following positions:

Administrative Assistant	USF Credit Union (credit union)	Germantown, TN
Patient Coordinator	Golden Eye Care (ophthalmologist office)	Memphis, TN
Bookkeeper	Function Junction (retail store)	Germantown, TN
Administrative Assistant	Mid-South, Inc. (manufacturer)	Germantown, TN
Insurance Coordinator	Foxx Brothers, Inc. (retail store)	Germantown, TN

EDUCATION

Oakland Community College, Auburn Hills, MI
Currently working toward accounting degree

Longview Community College, Memphis, TN
Completed liberal arts classes

Draughan's Business School
Memphis, TN
Diploma in general business

References Available Upon Request

MICHELLE P. HARTRICK

1728 Princess Lane
Farmington Hills, Michigan 48467
(313) 853-4754

Professional manager with a positive attitude and ability to supervise others; able to work well under pressure; computer proficient.

Professional Experience

Accounts Payable Department Manager, Autodyne Agency, Plymouth, MI — 1987-Present
(Available due to corporate-wide restructuring.)

Responsible for computerized and manual accounts payable at Components Group Disbursement Analysis center. Five corporate divisions are handled at this central accounts payable office.

Coordinate with engineers for payment approvals.
Certified proficient in EDS Dacor computer system.
Supervised 4 clerks who individually processed over 750 accounts payable invoices per month (average rate is 600 per month).

* Superior performance recognized by Management in the form of excellent reviews and regular increases.

Loan Counselor, United Mortgage Co., Livonia, MI — 1983 - 1987

Responsible for the collection of delinquent mortgages from 250-350 clients across the country per month.

Interviewed mortgagors to determine ability to reinstate mortgage.
Devised budget guidelines for mortgages.
Formulated delinquent client payment plan and monitored accounts continually.
Applied delinquent payments by computer to accounts.
Wrote personal and form letters to mortgagors demanding payment.
Notified VA, HUD and private insurance companies of default.
Coordinated field inspections for personal interviews when unable to contact mortgagors by telephone or mail.

* Achieved status of collector with the lowest delinquency rate through personal client contact and customer service.

Accounting Clerk, First of Michigan Trading Group, Royal Oak, MI — 1979 - 1982

Responsible for all accounts payable and accounts receivable. Calculated weekly bonuses and commissions based on productivity; paid account executives.

Set up new accounts based on commodity and market activity; maintained ongoing account records.
Manually computed and issued yearly profit and loss statements.
Computed margin calls when stock market prices fell.
Manually performed payroll including severance pay for 40 employees in 2 states.

* Commended by management for being a quick learner and dependable.

Office Manager/Bookkeeper, Decade Co., Flint, MI — 1973 - 1979

Responsible for managing company business office.
Calculated accounts payable and accounts receivable and posted journal entries.
Prepared accounts receivable statements; balanced with journal entries.
Computed payroll and all corporate taxes.

Developed catalogue of all parts, codes, and determined part inter-changeability.
Responsible for inventory control and the ordering of all parts and chemicals.
Calculated prices of products and services for all cash sales.
Worked with CPA as necessary.

Previous positions: Secretary/Receptionist, Clerk/Typist, Bookkeeper

PATRICIA LINCOLN

28934 Donald Court
Detroit, Michigan 48206
(313) 389 4930

CAREER SUMMARY:

In-depth diversified experience in accounting and financial planning, and financial analysis. Proficient with the following software packages: Lotus 1-2-3, Symphony & Harvard Graphics. Work on both PC's and Mainframes.

PROFESSIONAL EXPERIENCE:

Quarton Tire Company
Southfield, Michigan, 1965 - present

Financial Analyst — 1991 - Present
(Automotive Industry Group)

- Provided accounting and financial analysis support in the areas of sales reporting, operating expense control and capital spending fixed asset control
- Coordinated the development of annual $200,000 capital, operating expense ($10M) and net sales budgets of $928M.
- Developed, analyzed, and reported monthly forecasts of net sales, operating expense, and capital spending.
- Controlled physical assets, capital spending through appropriation request control and tracking physical inventory and asset record reconciliation, asset tagging and disposition control.

Financial Analyst — 1984 - 1991
(Original Equipment Research & Development Group)

- Developed operating budgets and project cost budgets of $65M for geographic locations; prepared management schedules and analysis for corporate review and approval.
- Developed monthly operating forecasts and detailed analyses of actual versus forecast/budgets for review at monthly staff meeting. Directed operating management in cost reduction/containment actions to meet budget objectives.
- Wrote and documented job procedures. Conducted training seminars for 220 employees to facilitate compliance with project/cost accounting.

  * First Project Leader to complete conversion of two separate systems into one single project cost reporting system following joint venture. Evaluated system alternatives, recommended selection and directed both MIS and user groups in implementation.
  * Active participant in annual Quality Circle Group which instituted cost avoidance/savings programs resulting in $350,000 annual savings. Received top corporate commendation.

Prior experience: Staff Accountant

EDUCATION

Associate of Applied Science, Accounting Major
Wayne County Community College

JACOB B. WITT

18264 Stanley Boulevard
Kalamazoo, Michigan 44852
(616) 489-2930

Extensive experience in Data Processing, the last 15 as an Administrative Support Specialist in a large IBM environment.

EMPLOYMENT HISTORY:

National Bank of Kalamazoo, 69711 Grand Avenue, Kalamazoo, Michigan

ADMINISTRATIVE SUPPORT SPECIALIST
Responsible for the review and analysis of business procedures and problems to coordinate the efforts of users and "systems" personnel. Confer with users to ascertain output requirements and then format for management reports. Liaison between organizational units and data processing department. Study existing data systems and recommend new systems to improve production and work flow. Review and approve documentation pertaining to new systems to meet current and projected needs. Supervise analysts.

- Maintained superior working relationship between "system" personnel and user department.

- Intricate knowledge of applications in the area of documentation review to include disaster recovery, report distribution and tape management.

- Worked with major areas such as:
  Stock Transfer
  Personal Trusts
  Mortgages
  Installment Loans
  Credit Card Applications
  Payroll

- Coordinated the conversion of a credit card company the bank purchased to insure the appropriate documentation consisting of 150-200 procedures.

Bank experience includes some programming and knowledge of UCC7-scheduling, telefile, hardware and JCL, CRT (3270) Terminal.

Previous Work Experience: Retail Sales, Data Entry, Bank Teller, Clerk, Customer Service

MILITARY EXPERIENCE:

Two years; Continental Army Command (CONARC), Ft. Madison, Maryland
Worked in data processing with IBM equipment.

EDUCATION AND TRAINING

Numerous in-house training programs to upgrade skills

B.S. - Computer Science
Indiana University
Bloomington, Indiana

ROXANNE N. LETTERMAN

21873 Timberidge
St. Cloud, California 97082
(805) 385-2932

Employment History
Homart Industries
Redman Automotive Division
Warren, California
1973 - Present
(available due to downsizing)

Electronic Data Processor
Responsible for development and processing of the salary and hourly payroll; utilizing labor and service, job rates, departmental changes and other pertinent data. Set up data controls, instruct operator as to which program to run utilizing tape controls. Participate in meetings with programmer, supervisor and department manager to establish work schedules for local plant and out of state locations.

* Developed payroll system procedures to coordinate efficient flow of jobs to be processed on a daily, weekly, monthly and year-end basis.

* Demonstrated a high proficiency with data entry that is recognized by management.

* Commended for exceptional ability to recognize, troubleshoot and solve complex problems.

Other Redman Assignments

Group leader of a five employee group involved in electronic data processing for the aftermarket.

Assembled and maintained technical manuals and mailed to customer. Maintained customer files.

Additional Experience

Owned and successfully operated retail business.

Education

Commerce High School
San Jose, California
Graduate

Redman Automotive
Trained in word processing
(in house)

Hobbies and Interests

Arts and crafts, Gardening

Volunteer Work

American Cancer Society
St Joseph's Hospital

References furnished upon request

BARTLEY H. EGGLESTON
3748 Windwood Drive
Dallas, Texas 46703
(305) 567-3354

CONSULTANT / EDUCATOR / INSTRUCTOR

An experienced systems analyst, manager and teacher of business/manufacturing systems.

MANUFACTURING SYSTEMS EXPERIENCE

Management Information Systems Manager — 1985 - Present
Orion Electronics Inc., United Motors — Dallas, TX

Responsible for managing all company computer resources in the design, development, and production of electronic systems for military and industrial applications (Discrete Manufacturing).

- Manager of department with annual budget of over $1.5 million.

- Chosen to implement comprehensive, innovative, all-inclusive MRP II Department of Defense Contract Management System consisting of engineering, material requirements planning, bills of material, purchasing, shop floor control, and numerous accounting systems. Installed new computer hardware system to handle the programs. All systems installed on time, on budget, and all performing as designed.

Manufacturing Systems Supervisor — 1975 - 1985
United Motors — Detroit, MI

Responsible for financial and material control systems for 27 remote manufacturing plants.

TEACHING EXPERIENCE

Business Data Processing Instructor — 1983 - 1985
Wayne County Community College — Detroit, MI

Taught Data Processing Systems Design, Basic, and Introduction to Data Processing courses, nights and Saturday while working full-time as Systems Supervisor.

SPECIAL TRAINING

Knowledgeable in MRP-II Systems, all basic business systems, including inventory and financial areas, and in barcode systems.

- Represented United and was chosen as Secretary of Barcode Standards Subcommittee, Automotive Industry Action Group of Detroit.

EDUCATION

Western Texas University — Pursuing Ed.D.
All course work completed

Western Michigan University — M.A.
Management

Central Texas University — B.S.
Mathematics

OTHER:
Southern Institute of Technology — 1990
Advanced Networking I (WANS)
Advanced Networking II (LANS)

KERRY A. CHUTE

1827 Southfield Road, Apt. #34C
Birmingham, Michigan 48009
(313) 645-8321 - Home

Over 15 years of experience in the data processing industry including software development, networking and operations. (Available due to corporate merger.)

Experience

PICA Corporation, Livonia, Michigan
Branch Manager 1985-1991

Manage FIGURES Product Line and Corporate MIS departments. Combined, these two organizations have a $2.7 million budget and 22 people.

The FIGURES Product Line Group develops and supports software systems in the financial arena with a target market of the telephone industry. This group's responsibilities also include training, documentation and market support.

The MIS department is responsible for providing software, hardware and consulting support for Accounting, Human Resources and field managers. This organization provides general management and capacity planning for TEXIC's VAX and PC hardware and wide and local area networks.

Project Leader 1983-1985

Managed project teams for development and support of database applications. Corporate-wide responsibility for technical support of System 1032 (DBMS) including contract negotiations, licensing and training.

XYZ Network Services, Bowling Green, Kentucky

Technical Consultant 1980-1983

Responsible for marketing and technical support for Value-Added Remarketing Group, an $6 million per year business unit. Major project was a two-year migration of clients from DECsystem-10 to VAX 11/785 based hardware. This included software staging, database loading and assisting network group in integration of ADP's packet, switched network to DEC's X.25 PSI software on the VAX.

Technical Specialist 1978-1980

Headquarters-based technical consultant for field. Responsible for trouble-shooting, application design and coordination, and impact assessment for major releases of internal and third party software. Also provided technical training and seminars for field staff.

Systems Analyst 1974-1977

Systems Analyst in the Programming Languages Department. Responsible for systems analysis and implementation, writing technical and user documentation, and consulting for field offices. Software projects included support of a Command Language Interpreter, development of an electronic mail system and host language interface for the programming languages. All development was on DECsystem-10 in FORTRAN and Assembler.

Education

B.S. Computer Science
Michigan State University, East Lansing, Michigan

DANIELLE T. CRAWFORD

4654 Drayton Avenue, Apt 3C
Troy, Ohio 53283
(419) 547-1353

Highly organized and skilled in computer-oriented administrative functions. Competencies include:

HARDWARE: IBM PS/2, IBM PC AT, IBM PC XT, AT&T, Decwriter

SOFTWARE: WordPerfect, Advanced Revelations, Lotus 1-2-3, Symphony, dBASE III, DisplayWrite 3, MultiMate, VideoShow, PictureIt

EXPERIENCE:

Whitman Food Products Associates, Troy, OH 1990-Present
(Available due to company downsizing.)

EXECUTIVE SECRETARY

Responsibilities include:

- Provide all administrative support for the following: Senior Vice President of Marketing, Vice President, Chief Financial Officer and President.
- Direct special projects which include gathering information for the Director of Personnel to report to parent company i.e., new hires, terminations, salary increases.
- Maintain personnel database - HRPRO, for 250 employees.
- Create and run reports on HRPRO for the Assistant Secretary/Treasurer who handles coordinating benefits for Whitman.
- Created capital expenditure format on Lotus for budgeting purposes
- Coordinate all travel arrangements
- Word processing and shorthand

Larix Sealing Systems, Troy, OH 1987-1990

MATERIALS COORDINATOR
Responsible for maintaining and updating material specification data base, working from notes prepared by engineers. Provided clerical support to Director of Materials and staff of six engineers. Utilized Wordperfect, Lotus 1-2-3 and Advanced Revelations.

- Developed and implemented database for engineering notices utilized by Marketing and Drafting Departments. This conversion from handwritten to computerized notices resulted in significant reduction in turnaround time.
- Developed and implemented database to track drawings. Resulted in improved efficiency in locating drawings and related documents.
- Performed data entry and check reconciliation for Accounting Department, utilizing IBM System/36.

Amtech Dealer Systems, Bloomfield, OH 1977-1987

ACCOUNTS RECEIVABLE COORDINATOR
Responsible for production of invoices for dealerships within Amtech. Calculated amount due on automobile dealers' test equipment, researched and applied taxes, and handled telephone inquiries from dealers. Prepared monthly summary of Accounts Receivable function for Corporate Headquarters. Utilized AT&T PC linked to IBM mainframe, in-house software, Symphony, and MultiMate.

ADMINISTRATIVE ASSISTANT
Responsible for clerical support for the Finance and Data Processing staff. Utilized IBM PCs using DisplayWrite 3, Lotus 1-2-3, VideoShow, PictureIt, and MultiMate.

COORDINATOR/SUPERVISOR (PURCHASING DEPARTMENT)
Responsible for supervising three people. Entered requisitions into the inventory control system, maintained a log of requisitions (using dBASE III). Interfaced with customers and vendors.

Meteor Office Services, Troy, OH 1973-1977

Worked various clerical positions such as receptionist, typist, file clerk, biller, and accounts receivable poster. As a Check Correlator responsible for printing checks, data entry, and computer backups.

EDUCATION

Oakdale County Community College, GPA 3.7
Currently working toward an Associates Degree - Business Administration

Courses: Computer and Data Processing Principles
Business BASIC Language
Shorthand
Principles of Accounting

MEMBERSHIPS

Central Ohio Business Group
Optimists Club of Troy

VOLUNTEER

Oakwood General Hospital Auxiliary
United Way

HOBBIES AND INTERESTS

Golf, Tennis and Gardening

WALLACE T. STEVENS

476 West Maine Avenue
Cabot Cove, Massachusetts 34726
(202) 478-0394

Experienced technical manager with strong employee relations skills. Capabilities in the following areas: Project Management, Data Communications, Systems Administration, and Training and Education. Extensive Mainframe, PC and LAN experience.

PROFESSIONAL HISTORY

PLANT SYSTEMS MANAGER
Hi-Mix Manufacturing, Inc., Westview, MA
1986 - Present, (available due to plant closing).

Managed the daily operation of all computer hardware and software at local facility and three remote locations. Responsible for the management of data communications, computer-related training and education, and systems utilization and security.

- Evaluated current technology and developed strategic plans to support long range corporate goals.
- Coordinated an ongoing program designed to provide continuous improvements in efficiency and ease of systems operation.
- Provided internal consultation to management regarding computer technologies.
- Directed and participated in the installation and modification of divisional systems and sub-systems.
- Served periodically on various plant task forces and teams involved in improving efficiency and quality.

DATA PROCESSING SUPERVISOR
C & C, Inc., West Grand Blvd., Westview, MA
1981- 1986

Supervised and directed data processing support for all system 36 daily, monthly and annual functions including: training operators, handling system failures and abnormal system events. Functioned as security officer, security administrator and contact for all on-site systems. Monitored system performance, installed and coordinated all upgrades. Responsible for set up, testing and maintaining communication configurations for two IBM system 36's.

Previous experience includes: Computer Support Technician, Computer Operator/Trouble-shooter

EDUCATION

Bachelor of Science
Computer Science

University of Indiana
Bloomington, IN

MEMBERSHIPS

Professional Association of Computer Managers - Board Member
American Society of Systems Managers

JOAN L. BELLANY

4637 Grove Street
Oak Park, Illinois 52039
(201) 382-3094

EMPLOYMENT HISTORY:

International Trucking, Inc. Chicago, Illinois
June 1975 - Present

SUPERVISOR OF DATA PROCESSING

Responsible for the management of all data processing personnel and operations for a major manufacturer of heavy duty trucks.

Supervised the hiring, development, training and evaluation of over 45 employees.

Developed departmental procedures insuring timely and accurate information needed by other department supervisors.

Initiated purchase of new equipment and systems in order to increase efficiency. Reviewed and evaluated hardware and a variety of software programs. Recommended and implemented new systems.

Managed an annual departmental budget of over $10 million.

* Commended by Management for reducing departmental overtime by 12%, and absenteeism by 7%.

* Outstanding performance reviews, excellent work and attendance records.

* Excellent people skills and a creative problem solver.

EDUCATION:

Northwestern University - Bachelor of Science
Computer Science and Accounting
Graduated with Honors

HOBBIES AND INTERESTS:

Sailing * Tennis

MEMBERSHIPS:

Professional Women's Association - Treasurer

Treasurer of PTA, Three years

References Upon Request

ANDREA B. JARVIS

1117 W. Burberry Drive
West Allis, Michigan 48332
(313) 783-9044

EMPLOYMENT HISTORY:

1976 - Present
U.S. Royal Stores, Inc.
15 U.S. Royal Plaza
Detroit, MI 48776

DATA ENTRY OPERATOR

Analyzed and solved problems that originated in keypunch concepts. Chosen to arrange layouts for quicker and easier keypunching. Responsible for training new employees. Strengthened the department by assuming more responsibility.

COMPUTER OPERATOR

Operated the IBM System 34 for over a year and operated the IBM Series 1800 for six months. Have basic knowledge of several programming languages.

Secretarial Skills:

Developed typing skills by taking two years of typing in high school.

Presently improving skills by completing advanced typing class in college. Type 75 wpm with high degree of accuracy.

EDUCATION:

Chattahoochee Valley Community College — Secretarial Courses, Accounting/Word Processing

Career Training Institute
Career Training Institute — Keypunch Course

Bay City High School
Bay City, MI — Graduated
Business Courses

HOBBIES & INTERESTS: Reading, sewing and gardening

PERSONAL DATA: Married, excellent health
Willing to relocate

TONY J. McMANN

3800 W. County Rd. 17 | Grayfield, Washington 98103 | (303) 912-9920

SKILLS AND ABILITIES:

* Well-organized professional with demonstrated strength in working independently and as a member of a team to accomplish established goals.
* Organization and completion of detailed communication projects.
* Time management skills; ability to work under pressure and deadlines.
* Accurate and efficient background and in-person research.
* Skilled in assisting others to grasp the substance and implications of new subjects and perspectives.
* Excellent interpersonal and written communication skills.

PROFESSIONAL HISTORY:

EDITOR — 1991
The Real Estate INSIDER (monthly publication), 900 Madison, Eugene, OR 98104
(Available due to corporate downsizing.)

Produced editorial content of new magazine concerned with commercial real estate in the Puget Sound market. Acquired background in real estate issues through extensive research in the field. Interviewed professionals involved in brokerage, development, governmental and economic development activities. Recruited and directed writing of contributing authors; edited all articles published. Authored news stories, features and editorials. Coordinated all non-advertising aspects of publication production.

EDITOR/COMPLIANCE OFFICER — 1984-1986
All Four Seasons Resorts, Inc., 3440 12th Ave., NE, Suite 600, Redmond, WA 98053

Elicited general interest information from management, staff and customers. Generated a variety of articles for monthly member tabloid. Edited all copy; supervised production. Interpreted statutes and rules regulating the company's sales, marketing, contractual and administrative activities. Prescribed, organized and monitored company response to regulatory requirements in seven states.

ACCOUNT EXECUTIVE — 1982-1984
Kearns Willard, Inc. (executive search firm) 1525 Fourth Ave., Suite 610, Seattle WA 98101

Business development and personnel recruiting. Generated clientele in the banking industry. Located candidates to fill executive level banking positions.

LEGISLATIVE ANALYST — 1975-1982
State of Indiana, House of Representatives

Translated statute into understandable English; prepared oral and written reports, briefed legislators on content of and controversy concerning proposed law. Communicated daily with legislators, lobbyists and state departmental executives.

GRADUATE TEACHING ASSISTANT — 1971-1975
Department of Philosophy, Indiana University, Bloomington, IN 45874

EDUCATION: Indiana University
M.A., Philosophy
B.A., Philosophy (Phi Beta Kappa)

LIONEL J. PENNING

29382 Tiffany
Westland, Michigan 48389
(313) 364-2134

Extensive experience in positions of increasing responsibility for manufacturing engineering with Sterling Corporation, including production engineering, manufacturing feasibility, cost reduction and quality improvement. As Advanced Manufacturing Feasibility Engineering Manager, had corporate responsibility for determining the manufacturing feasibility of the Prince Spirit and Shadow, Liberty Acclaim and Sundance and Sterling LeBaron car lines. Accomplished objectives through a combination of perseverance, technical skills and management ability.

PROFESSIONAL HISTORY:

Sterling Corporation, Westland, Michigan (available due to plant closing.)
1973 - Present

ACCOMPLISHMENTS

Planning and Organizing

. Led a manufacturing team in determining manufacturing feasibility of Sterling Corporation's 1989 "Z" body.

. Coordinated launch for paint, trim, chassis and final areas from 1974 through 1980.

. Developed method of assembly and built pilot vehicles for Prince Truck Operations including first extended van, Ramcharger, medium duty tilt cab and light duty pickup.

Cost Control

. Achieved at least $25.00 per vehicle cost avoidance on grille installation for Prince Shadow and Liberty Sundance.

. Reduced front end assembly cost on Prince Spirit and Liberty Acclaim by innovative processing.

. Initiated program to control reproduction service cost for Prince Truck Operations Production Engineering.

Communication

. Developed and presented vice presidential-level New Model Product Reviews.

. Developed and presented weekly New Model Update information.

. Coordinated the interfacing of manufacturing with engineering and styling on new products.

Analysis

. Interpreted product designs for manufacturing feasibility.

. Contributed to improvements for locating steering columns, instrument panels and tail lamps.

Innovation

. Initiated method and sequence of assembly for front end of Prince Spirit and Liberty Acclaim resulting with patent on headlamp locating clip (co-inventor).

. Invented self-locating bracket for installation of loose upper deck panel.

EDUCATION

Mechanical Engineering, University of Toledo
Engineering Computer Principles, Sterling Institute
Journeyman Tool and Die Maker, Sterling Apprenticeship Program
Numerous management and technical company training programs

DONALD ALLEN JAMES

Rt #2 Box 94B
Davenport, Iowa 78354
(425) 463-2467

EMPLOYMENT HISTORY

Acme Motors - Fenster, Iowa — 1974 to Present
(Available due to plant closing.)

GROUP LEADER - INDUSTRIAL ENGINEERING

Over 15 years of supervisory experience for a major world-wide automobile manufacturer. Included in this experience is 12 years of Industrial Engineering activities in the Trim section of the assembly line. Supervisory experience includes 55 hourly employees on the assembly line in prior years and 375 hourly employees in the Trim area presently. Responsible for establishing the most cost effective methods of utilizing available manpower, determining line speeds and conducting time and motion studies.

- Redesigned product line, reducing man hours and material cost by $155,000.
- Excellent working rapport with all levels of management, hourly employees and union representatives.
- Excellent work and attendance records.
- Outstanding job performance reviews for the past 10 years.
- Excellent people skills and problem solving.

Previous experience includes:

Machine Operator, Inventory Control Clerk, Warehousing, Assembly and Fork-lift Driver

EDUCATION

Draughts College, Davenport, Iowa
B.S. in Business Administration

Acme Institute
Courses in management, quality, time study
and computerized data recovery systems.

MILITARY SERVICE

U. S. MARINE CORPS - Honorable Discharge

HOBBIES AND INTERESTS

Hunting, fishing, golf and traveling

VOLUNTEER WORK

School Board Member, Groves High School Booster Club
Little League Baseball team.

TERRY T. THOMAS

32748 East 13th Street
Parkvale, New Jersey 08511
(609) 389-4903

PROFESSIONAL EXPERIENCE:

ENGINEERING MANAGER — May/83 to Present
Ex-press-o Corporation, Lemax Cam operation, Parkvale, New Jersey

Responsible for Product Design and Development of Computer Aided Manufacturing (CAM) systems, Numerical Control Punched Tape and Interface products for the Machine Tool Industry. Systems include: Distributed Numerical Control (DNC), Machine Monitoring and Shop Floor Data Entry systems, Computer Integrated Manufacturing (CIM) and Artificial Intelligence (AI).

Staffed and equipped engineering department to support business conversion from hardware-oriented organization to a system supplier and integrator. Developed product designs and marketing strategies in support of business plan objectives.

Served secondary position as Marketing Manager for the seven state western region while a National Representative Organization was developed. Participated in the selection and training of this organization.

Wrote DNC tutorial and computerized cost justification programs for Marketing. These programs were used in a major advertising campaign which created the majority of the leads resulting in product sales. Actively participated in trade shows, customer negotiations, system quote and integration activities.

GROUP SUPERVISOR OF PRODUCT ENGINEERING — May/78 to May/83
BBCO Incorporated, New York City, New York

Supervised Computer Peripheral, Avionics and Video Engineering groups. Developed Numerical Control Tape Readers and Punches for Machine Tool Industry, Audio Entertainment and Passenger Service Systems for Commercial Airline Industry, Time Code Generation Queing and Video Editing Systems for Broadcast Industry.

Served as Program Manager on BBCO's largest single contract for Airborne Audio/Passenger Service systems. Follow-up contracts from program success exceeded the original $2.4 million contract. An avionics product group was developed from this venture.

Successfully revitalized the Peripheral products group through new product offerings and restructuring which resulted in profitable sale with royalty incentives to Ex-press-o Corporation.

ADDITIONAL PRODUCT DESIGN ACHIEVEMENTS:

Cargo and Baggage Handling Vehicles plus Terminal Systems for the Airline Industry. Material Handling Vehicles and specialized systems for industrial applications.

Life Support and Flight Safety equipment for Military Pilots.

EDUCATION:

Chicago Technical College - Mechanical Engineering
Detroit Technical College - Program Management
New York State University - Software Design Techniques
University of New York - Behavioral Science for Managers

ORGANIZATIONAL AFFILIATIONS:

Association for Integrated Manufacturing
Manufacturing Engineers Society
Society of Automation Protocol (SAP) users group

PAPERS AUTHORED AND PRESENTED:

"Conductivity of Water"
USAF and Military Service

"A New Solution - Distributed Numerical Control and Other Systems"
TRF Flexible Machining Centers Clinic, Chicago, Illinois

"A Building Block Approach to DNC"
Advanced Manufacturing Systems Conference, Detroit, Michigan

PUBLICATIONS:

"A Modular Approach to DNC"
Journal of Manufacturing Technology

"A Systems Approach to DNC"
Machine Tools East magazine - Nov. 1986

PATENTS:

3,989,809 - Trigger Release Circuit
4,092,928 - Auto Harness Assembly

ORGANIZATIONAL AFFILIATIONS:

*Association for Integrated Manufacturing

*Manufacturing Engineers Society

*Society of Automation Protocol (SAP) users group

EDUCATION:

Chicago Technical College - Mechanical Engineering
Detroit Technical College - Program Management
New York State University - Software Design Techniques
New York State - Supervisory Development Program
University of New York - Behavioral Science for Managers
Plus numerous courses, seminars and studies to support position requirements.

PERSONAL:

Married, Veteran, Excellent health

HOBBIES:

Golf, Jogging, Painting, and Sailing

References available on request

MICHAEL L. SIMMONS

15178 Pettigrew Drive
Rowing, Texas 84068
(414) 812-5223

Employment History

American Manufacturing Co. 3434 South Houston Place, Galton, TX 85677
1973 - 1991, (available due to plant closing).

INDUSTRIAL ENGINEER

Member of the engineering team that developed and applied the most cost effective sequences and methods to the production operations. Plans were developed based upon existing equipment, facility, scheduled time allowed and tooling requirements. Developed and applied standard time data for cost control. Checked and verified blueprints. Wrote input sheets for computer-generated work routing.

Previous duties and responsibilities included:

SHOP CONTROL SUPERVISOR

Organized, scheduled and released Structural Shop work load, based on master completion requirements. Insured timely completion of work orders by effective communication with all levels of factory supervision. Reported on status of work orders in progress to Production Manager. Supervised 18 Expediters and Material Handlers.

* Outstanding attendance and safety records.

* Commended by Management for cost reduction ideas which generated $53,748 in savings.

* Developed a training program for Engineering Technicians in time and motion study.

Education

J.R. Tandy High School   Union City, TX
Graduated   College Prep.

Safety & Supervisor Training, 1976
O.S.H.A. & American In-House Training

Personal Data

Married Two Children Excellent Health
Willing to Travel or Relocate

References Available Upon Request

TYLER S. SHORHAM

38757 South Creek Drive
Holly, Michigan 49064
(313) 537-6449

Hands-on manager with broad automotive engineering and manufacturing experience combined with excellent technical and interpersonal skills. Strengths in problem solving, training and development of engineering and technical personnel, planning, and program management.

Expertise in Body and Chassis Design and Development, Vehicle Dynamics, Product and Quality Engineering, Service, and the vehicle manufacturing process.

ACCOMPLISHMENTS

* Developed and led on-site cross functional new model launch task forces at pilot and assembly plants, reducing or eliminating the number of launch problems or concerns
* Coordinated engineering assistance and training for General Motors assembly plants and kept them up-to-date on engineering changes and advanced product designs
* Initiated, developed, and implemented a Process Control Department for a General Motors assembly plant
* Supervised Quality Control Departments, Layout Inspection, and Product Engineering Departments resolving manufacturing problems and meeting all corporate goals
* Developed and published a Port operations guide and a Police and Fleet special equipment parts book for General Motors Service & Parts Division.
* Participated in the creation and revision of corporate process and performance standards
* Developed and implemented solutions to major chassis design and vehicle component assembly problems averting lost time and vehicle quality or durability problems
* Graduated from the General Motors Institute of Engineering. Evaluated, sponsored and taught new students for the General Motors Institute of Engineering
* Construction, demolition, and unit command experience as a commissioned officer in the U.S. Army Corps of Engineers

PROFESSIONAL EXPERIENCE

General Motors Corporation — 1956-1991

Launch Program Executive, Product Engineering Supervisor, Process Control Manager, Vehicle Serviceability Specialist, Service Development Engineer, Contact Engineering, Supervisor, Senior Quality Engineer, Senior Product Development Engineer, Product Development Engineer, Test and Development Engineer, Student Engineer

EDUCATION

General Motors Institute — M.A.E.
Ohio State University — B.S.M.E.

ADDITIONAL TRAINING

Process Redesign
Statistical Process Control
Supervisor Training

General Motors Quality Education System
Geometric Dimensioning & Tolerancing
Interaction Management

ROGER M. JOHNSON

547 West Maple Road
Mill Valley, New Jersey 08513
(609) 489-3094

Over fifteen years of professional experience with a major manufacturer of snack products.

PROFESSIONAL HISTORY:

International Snacks, Inc., Mill Valley, New Jersey
June 1973 - Present (available due to corporate downsizing).

SENIOR PROJECT ENGINEER
Responsible for supervising, planning, coordinating and directing the activities of an engineering group involved in multiple research and development projects.

Independently designed and conducted experiments to meet research objectives. Capable of identifying, planning and executing projects with minimal supervision.

Responsible for training of engineers at various levels, and multi-project management across multi-disciplines. Demonstrated project management principles.

Utilized a wide range of engineering principles to problem solve. Developed a broad technical knowledge base which permits refined troubleshooting.

Maintained and continuously improved upon technical competency.

PROJECT ENGINEER
Independently designed and conducted experiments to meet research objectives. Identified resource needs to accomplish project objectives. Interpreted experimental results and identified follow-up strategies.

Successfully worked within stated budget parameters. Managed $1.7 million annually.

Effectively utilized internal and external resources to achieve objectives.

Demonstrated initiative in seeking and applying information from outside sources to conduct research.

Responsible for safety, maintenance and daily operation of assigned laboratory and pilot plant equipment.

EDUCATION:

M.S. - University of Rhode Island, Chemical Engineering
B.S. - New York University

PUBLICATIONS AND PRESENTATIONS:

Food Science Journal, 1981
"Questions on Additives"

"Questions on Additives" presented before the American Association of Food Researchers, Atlanta, Georgia - 1982.

TAYLOR M. LONGTREE

39480 15 Mile Road Farmington, Michigan 48047 (313) 674-9742

Experienced Project Engineer with over 20 years of background in plastics from the plant floor to plant management. A self-motivated team player, familiar with prototype and production injection molds, support equipment and processing of a wide variety of materials.

ACCOMPLISHMENTS:

- Placed and followed all tooling for program from prototype thru production.
- Coordinated design, tool build and sampling of tools.
- Attended customer meetings and reported progress. Corporate offices and plants were out of state and had to be kept updated. This resulted in an on-time launch within the specified objectives.

PROFESSIONAL HISTORY

SENIOR TOOL ENGINEER 1988-1991
ABC Vehicle Safety Systems, Farmington, MI
(Available due to reduction in work force.)
Member of a team involved in the design and tooling of all seat belt parts.

PROJECT ENGINEER 1984-1988
O'Halloran Engineering Corporation, Dearfield, MI
Responsible for reporting progress on all tooling for all programs.
Reviewed design and tooling for largest interior trim project outsourced by Ford Motor Co.

PROJECT ENGINEER 1983-1984
Dimoaplast Corp., Taylor, MI
(Bankrupt)
Customer contact and follow-up for design and tooling.
Designed gauges, fixtures. Directed product design.

MANAGER OF TOOL ENGINEERING 1979-1983
Bendix Safety Restraints, Inc., Mt. Clemens, MI
(Reduction in work force.)
Supervised staff of six including engineers, lab technician and clerk.
Placed and followed all tooling, both plastic and stampling.

PLANT MANAGER 1975-1979
Service Plastics, Livonia, MI
(Plant sold)
Responsible for day-to-day operations for mold try-out speciality company.
Hired, supervised and provided training for employees and temporaries.
Managed 7-person permanent staff.

EDUCATION: Journeyman's Card - Federal Certificate of Completion

4-Yr Tool and Die Apprenticeship - Manufacturing Trades

Graduated - Cass Technical School, Detroit, MI

MEMBERSHIPS: Society of Plastics Engineers
Elks

THOMAS S. HOMERSKI

22324 Red River Run
New Delaware, Indiana 49711
(219) 354-5528

Over 20 years of experience in manufacturing management with in-depth background in production processes and product quality. Extensive experience in the start-up of new facilities. Ability to conceptualize and deliver new methods to increase quality, productivity and reduce costs. Proficient at developing and motivating a productive staff. Significant union/management relations experience. Goal-oriented with excellent interpersonal, planning, and organizational skills.

PROFESSIONAL EXPERIENCE:

STUDEBAKER MOTOR COMPANY 1979 - Present

Director Quality and Product Engineering - Highland Park, IN 1989-present

Coordinate resolution of engineering and quality issues between 4 assembly plants, Corporate Engineering, and Supplier Quality Assurance (SQA). Perform process conformance audits at assembly plants. Monitor and investigate customer warranty complaints for correction. Instrumental in obtaining engineering and supplier quality changes that have resulted in a 12% quality improvement during the first six months of production.

Manager, Process Development
Mick Avenue Process Development Center - South Bend, IN 1987-1989

Directed and coordinated the installation of the latest technology in automobile manufacturing process equipment and machinery. This included working with equipment suppliers and platform directors on the installation and tryout of robots, conveyor systems, and weld equipment for current and future assembly processes. Managed 15 people.

Planned the manufacturing process for a low volume sports car to begin production in December 1992. As a result of overall efforts, various equipment will be utilized at the new Studebaker Jesperson Avenue Plant.

Plant Manager - Newark, NJ 1983-1987

Managed an automotive assembly plant that produced 750 automobiles per day on two shifts. Directed a staff of 15, as well as 150 indirect salaried employees and 3,500 indirect hourly employees.

* Improved first-time capability (FTC) from 45% to 72%.
* Implemented Just-In-Time material delivery and sequence part delivery to reduce inventory.
* Established block painting resulted in approximately $1 million in annual savings.
* Reduced inspection and repair by 15%.
* Increased productivity as measured by vehicles produced per employee by 20% from 1983to 1987.
* Installed over $12 million of plant improvements without production interruption.

Plant Manager, Studebaker Motors Pilot Plant - South Bend, IN 1982-1983

This plant performed pre-production activities of future products. Managed and developed the production process and tooling requirements for new vehicles prior to assembly plant production.

Operations Manager, Studebaker Newark Assembly Plant - Newark, NJ 1980-1982

Directed 2nd shift activities.

Quality Manager, Diversified Operations Division - South Bend, IN 1978-1980

Directed the quality systems for 12 diversified operations component plants, including an assembly plant in Mexico.

Developed a quality control system within the Studebaker DeMexico car and truck assembly plants which resulted in parity between Mexican and U.S. standards, thus permitting import of Mexican built automobiles to the U.S.

STERLING MOTORS OF AMERICA 1972-1978

General Superintendent of Quality Control - Stanton, IL 1975-1978

Directed all phases of the Quality Control Department, including outgoing product quality.

Superintendent of Product Engineering - Stanton, IL 1972-1975

Established the Product Engineering Department as part of the plant start-up of operations. This included development of department procedures and vehicle specification manuals.

Progressive Engineering Positions 1965-1972

EDUCATION/TRAINING:

Northland Institute, Midland, IN
B.S. in Management

Allied Institute of Technology, Chicago, IL
Certificate in Mechanical Drafting

Numerous courses and seminars including:

Crosby - Quality Improvement Process, Statistical Process Control,
Failure Mode Effect analysis, Quality Functional Development,
American Supplier Institute - Two week study in Japan of Total Quality Control

MAXWELL J. SMARDISH

481315 South Cherry Bend Drive
Bloomfield Hills, Michigan 49038
Office: 313/578-4545
Home: 313/776-9789

EXPERIENCE:

Beaman-Unyrich Tire Company, 1969 - Present
DIRECTOR OF RESEARCH AND DEVELOPMENT, 1986 - Present

Responsible for all materials used in O.E. tires (56 employees). Strategically plan and coordinate all resources for materials R&D. Manage Chemical, Physical, and Textile Laboratories.

- Improved department efficiency and maintained goals while staff was reduced by 40% and budget was reduced by 60%.
- Designed continuing training programs which significantly enhanced employees' capabilities.
- Developed state-of-art materials for meeting O.E. tire requirements.
- Achieved preeminence in applied materials technology through the introduction of several new state-of-art materials to original equipment tires - significantly improving performance.
- Provide technical assistance to the producing plants.
- Improved product quality while saving in excess of $20 million in materials and processing costs.
- Established materials and processing specifications; approve new and alternate sources of raw materials.
- Achieved better response from suppliers to company needs through closer relationships resulting in many industry firsts.

MANAGER, MATERIALS COMPOUNDING AND PROCESSING, 1980 - 1986

Responsible for research, development, evaluation, and processing of all rubber materials used in tires. Provided compounds which met customer requirements as well as being acceptable with respect to quality, cost and manufacturing. Supervise technical group of 15 B.S. and Ph.D. Engineers and Chemists.

- Developed first corporate computerized interactive recipe management system resulting in 80% time savings.
- Designed data acquisition systems for laboratory instrumentation resulting in at least 55% time savings and greatly improved reporting.
- Implemented first computer-based training in tire technology.

RESEARCH SCIENTIST, Tire Division, 1975 - 1980

Research, development, and evaluation of tire compounds and reinforcing materials, including supervisory duties.

- Developed first tire compounds specifically designed for low fuel consumption.

RESEARCH SCIENTIST, Corporate Research Center, 1969 - 1975

Research in textiles, specialty polymers, and tire compounds. Wrote proposals for, received, and carried out government contract research in pollution control.

Prior positions as : Research Assistant, Technician, Intern

EDUCATION:

The University of Michigan
Ann Arbor, Michigan
Postdoctoral Fellow
Nuclear magnetic resonance applied to deuterium and hydrogen in the solid and liquid states.

Michigan State University
East Lansing, Michigan
Ph.D., Physical Chemistry, (minors in mathematics and Physics)
N.A.S.A. Scholarship

Indiana University
Bloomington, Indiana
M.S., Physical Chemistry (minor in Metallurgy)
Cobalt-magnesium phase relationships

Iowa State University
Ames, Ohio
B.S., Chemistry (minor in Mathematics)
Received "Outstanding Senior in Chemistry" award from American Institute of Chemists

PUBLICATIONS AND AFFILIATIONS:

Seventeen publications in Metallurgy, Solid State Physics, Pollution Control, and Polymer Physics. Authored hundreds of technical documents.

American Chemical Society
Society of Automotive Engineers
National Association of Technical Executives

JOHN C. RIVERS

12345 Someplace Avenue
Anytown, Ohio 54908
(312) 648-9876

Over twenty years of diversified management experience in the beverage industry with Johnson Bottling and Distribution Company, Columbus, Ohio.

Professional History:

DIRECTOR OF FINANCE ADMINISTRATION

Directed and coordinated administrative and financial activities of the Operations Department including: capital expenditure budgets, operating budgets, productivity improvement and manpower planning, organization and staffing.

-Enhanced profitability by recommending and implementing budget reductions and establishing close control over expenditures.

-Reduced projected operating budget costs over $100,000 per year with an objectives program.

WAREHOUSING AND SHIPPING MANAGER

Directed all warehousing, shipping and receiving functions for a high-speed packaging operation, producing 500,000 cases per day, servicing over 375 independent distributors in a sixteen-state area. Responsible for over 250 unionized employees.

-Reduced manpower requirements by 50% and increased direct loading 40% by establishing a centralized operation.

-Improved customer service and reduced costs by establishing a pre-load and drop trailer shipping program.

MANAGER OF PRODUCTION CONTROL

Planned and scheduled production line utilization, finished goods inventories and manpower requirements for 10 million barrel packaging plant. Minimized manning and labor turnover using production control techniques. Coordinated computerization of daily labor and cost reporting systems.

Education:

University of Michigan	B.S., Business Administration
Ann Arbor, Michigan	Major in Accounting

Memberships:

National Association of Finance Administrators
Rotary International

Hobbies:

Swimming, Tennis and Golf

MARK P. SMITH

14760 Barnes Avenue
Anytown, Anystate 45987
(211) 760-0702

Twenty years' experience in sales, manufacturing and administration, including full profit center responsibility.

ADMINISTRATION

PRESIDENT - Managed a fully integrated truck manufacturing operation with complete profit/loss responsibility. During this period:

* Expanded gross sales from $7M to $88M annually.
* Achieved annual unit sales growth from 380 to 2,600.
* Doubled percentage of market penetration.
* Supervised growth of workforce from 235 to 820.
* Decreased unit costs by 10% despite inflation.
* Maintained excellent employee relations - no work stoppage in ten years.

MANUFACTURING

PLANT MANAGER - Director of manufacturing and manufacturing support operations for plant producing 1,700 trucks annually.

* Planned and implemented three plant expansions, which increased plant size from 20,000 to 140,000 sq. ft.
* Reduced assembly hours by 36%.

SALES/MARKETING

GENERAL SALES MANAGER - Developed and implemented nationwide sales program.

* Maintained liaison with 720 dealers to resolve problems and gather market and customer preference information.
* Directed follow-up and resolution of customer problems.
* Developed successful product advertising strategy.

WORK HISTORY:

Anytown Truck Manufacturing Corp., Anytown, AN 1971 - 1991

EDUCATION:

Harmond University - Advanced Management Program
Oregon State University B.A. - Business Administration

ALEXANDER D. REISSMAN

3858 Sterling Avenue
St. Louis, Missouri 67085
(903) 564-3030

Twenty years' experience in business unit management, marketing, product planning, training, and sales, including full profit center responsibility.

Professional Experience

DIRECTOR
Venus, a subsidiary of Techno Data Systems, St. Louis, MO, 1984-Present.
Managed kit computers, measurement instruments, and product planning for educational products. Computer products ranged from 8088 to 80386/25 based desktops and laptops. During this period:

*Brought 65 new products to the market, an average of 1.5 products per month. 75% of these products were financially successful.

*Supervised up to 15 subordinates and managed budgets up to $1.5 million.

*Developed five-year product development plan for education group and strategic plans for computers and instruments.

*Maintained excellent employee relations--no voluntary staff turnover.

*Frequently traveled to Far East to select products and supervise development.

MARKETING MANAGER
Icor, Corporation, Minden, Kansas City, MO, 1978-1984.
Located and trained distributors, sold product, created collateral literature, set prices, negotiated OEM arrangements, provided business planning and interfaced with financial backers. Icor was a start-up magnetic video disk manufacturer.

MANAGER OF PRODUCT PROMOTION AND MARKET DEVELOPMENT
Barkley Iowa Corp., Kansas City, MO, 1976-1978.
Responsible for a variety of duties for a manufacturer of monitoring equipment for large machines.

*Supervised advertising, product management, pricing and OEM contract administration.

*Reduced staff from twenty to sixteen and dramatically increased quantity and quality of department output.

*Managed budgets up to $750,000 per year, always meeting budget.

*Performed quantitative market research to help fine-tune company's mix of price, service level, attention to quality, etc.

MARKETING PRODUCT MANAGER
Elec-tronix, Denver, Colorado, 1972-1976.
Responsible for product planning, new product introduction, sales support, pricing, and promotional activities for XY and video monitors used in medical diagnostic imaging, scientific instrumentation, and measurement equipment.

*Helped negotiate million-dollar contracts with OEM customers.

*Participated in building business unit revenue by 150% in three years.

*Participated in keeping business unit profitable during this growth.

FIELD ENGINEER TRAINING MANAGER
Managed the Elec-tronix Field Engineer Training program. Supervised 3-4 instructors and personally did the sales training.

FIELD ENGINEER (salesman) Sold Elec-tronix products.

Previous experience:

INSTRUCTOR--University of Colorado, Boulder, Co. Instructed upper division classes in Marketing Channels, Industrial Marketing, Production Management, and Quantitative Methods. Received very favorable student ratings.

EDUCATION

MBA, University of Missouri, Kansas City, MO

BS - Physics, Western Michigan University, Kalamazoo, Michigan

CHRISTINA J. BATTON

4839 East 54th Street
Grand Junction, Colorado 89356
(791) 7567-2930

A highly professional Radiologic technologist with MRI and supervisory skills.

EMPLOYMENT:

Mott-Southeastern Hospital, Grand Junction, CO **1974-1990**

Magnetic Resonance Imaging Staff Technologist 1983-1990

Responsibilities:

* Ran patient schedule, performing studies and adjusting for emergencies.
* Performed patient care, including assisting nurses and doctors in starting IVs and reassuring anxious patients during scans.
* Coordinated the requests of the ordering physicians and staff physicians.
* Assisted the supervisor with ordering supplies and acted as control tech for four staff members in his absence.
* Operated computer to track patients and retrieve reports.
* Assisted in training new technologists and x-ray students in basic principles of physics, anatomy, and computer operations.
* Ran troubleshooting procedures to determine sources of malfunction and appropriate measures to correct.
* Filled in at the receptionist desk when short staffed.

Accomplishments:

* Suggested a communication system which was implemented throughout the entire radiology department.
* Proposed front desk reorganization assigning specific responsibilities to each individual.
* Discovered a short-cut to reset the computer automatically when a certain error occurred in the software.
* Worked with film quality assurance to develop a reporting form which revealed errors and the percentage of film discarded. Technologists' errors were identified and retraining reduced occurrences.
* Participated in the development of the department from its inception, suggesting scheduling and management adjustments which resulted in smoother operations.

CT Staff Technologist/Staff Supervisor Diagnostic Imaging 1979-1983

Accomplishments:

* Selected to perform the most difficult procedures because of record of accuracy.
* Chosen to participate in a test of 3M equipment.
* Appointed evening shift staff supervisor in charge of seven staff members.

Previous Employment included: Nurse Aide, Ward Clerk, Housekeeping Supervisor

EDUCATION:

Radiologic Technologist Certificate, Bald Mountain Pass Hospital

Lincoln High School, Bald Mountain Pass, Minnesota

HENRIETTA K. LOCKWOOD

Summit View Apartments
3985 Mount Hood Drive, Apt. 34
Rochester Hills, Michigan 49045
(313) 387-5903

Registered Nurse with over 15 years' experience in Labor and Delivery with strong interest in Childbirth Education.

PROFESSIONAL EXPERIENCE

REGISTERED NURSE
Henry Ford Hospital, Detroit, Michigan 1984 - Present

Proficient in traditional labor and delivery setting, LDR concept and outpatient birth center services.

Duties included:

Prepare couples for childbirth through Birth and Family Education Department. (Private classes also offered.)
Evaluate labor and admit clients into Birthing Center.
Assess labor progress and provide nursing care to women throughout labor, delivery and postpartum.
Provide nursing care to infants.
Assist with breast-feeding.
Prepare patients for an early discharge home.
Initiate follow-up phone calls to families during postpartum.

STAFF NURSE
Labor and Delivery, St. Mary's Hospital, Terre Haute, Indiana 1978 - 1983

OFFICE NURSE
Dr. D. J. Chang, Obstetrician and Gynecologist, Ft. Wayne Indiana 1977 - 1978

STAFF NURSE
Labor and Delivery, St. Peter's Hospital, South Bend, Indiana 1975 - 1976

STAFF NURSE
Labor and Delivery, St. Joseph's Hospital, Toledo, Ohio 1974 - 1975

Previous experience includes: Nurse Aide, Emergency Room Nurse, Home Health Aide

EDUCATION

Received N.I.C.L.S. (Newborn, Infant, Child Life Support) Certification from Henry Ford Hospital, Detroit, Michigan, 1987

Received A.S.P.O. (American Society for Psychoprophylaxis in Obstetrics) Certification in Childbirth Education, 1985

Northern Michigan College of Nursing
Marquette, Michigan
Bachelor of Science in Nursing

References Available Upon Request

STANLEY B. MCDUFF

38477 Peachtree Lane
Dayton, Ohio 38955
(374) 394-3950

PROFESSIONAL EXPERIENCE

Chief Pharmacist 1987-Present
St. Mary's Hospital, Dayton, OH
Managed the pharmacy department in a profitable and efficient manner through inventory control, purchasing contracts, manpower studies (F.T.E.'s vs. output), and quality assurance studies. Hired, supervised and terminated employees. Established policies and procedures. Active on Pharmacy and Therapeutics Committee.

Pharmacy Manager 1984-1987
Wyman Foods (major regional chain), Dayton, OH
Responsible for the overall management of the pharmacy department. Includes subordinate supervision, inventory control, customer service promotion, product selection and display, public relations work with area physicians, and community wellness programs.

Launched two new pharmacy departments - designed physical layout, assembled initial inventory and implemented promotions. Sales volume increased an average of 15% monthly.

Applied cost benefit analysis to drug product selection to both maximize management profit and minimize costs to the customer.

Decreased on-hand inventory by 20% through sales analysis.

Staff Pharmacist 1979-1984
Indiana Medical Center, Ft. Wayne, IN
Assisted in implementing an inpatient unit dose dispensing system including bringing on-line a pharmacy computer software system. Responsible for drug information, technician supervision, and inpatient dispensing.

Staff Pharmacist
Willow Run Medical Center, Columbus, OH 1972-1979
Instituted and managed an inpatient/outpatient oncology chemotherapy program, including billing and purchasing. Formulated and performed quality assurance studies. Assisted in implementing a unit-dose dispensing system. Advised and participated in a reorganization of the pharmacy's physical layout.

EDUCATION

B.S., Pharmacy,
Ohio State University

HONORS

Beta Gamma Sigma - honorary business fraternity

HOBBIES

Investments, Racquetball, Softball, Outdoor Activities

PERSONAL

Married * One child * Health Excellent

Charlene C. Romano

385 Maple Court
Comstock, New York 20389
(527) 635-6022

Professional Experience

REGISTERED NURSE
United Nursing, Inc., Comstock, New York 1982-1991

Assigned to various areas including Emergency Department, OR (Holding, Recovery, One-Day Surgery), Surgical and Progressive Care Units in local hospital. Part-time member of administrative staff, responsible for appraisals of nursing staff. Coordinated and implemented AHA Basic Life Support - CPR program.

ADJUNCT CLINICAL INSTRUCTOR
Lane County College, New York, NY 1979-1982

Responsibilities included teaching, supervising and evaluating registered nursing students on the medical, surgical and PCU units at local hospitals. Co-taught laboratory skills course.

CLINICAL COORDINATOR
Lifeline Nursing, Inc., New York, NY 1978-1979

Responsible for interviewing and hiring of employees, evaluation and assessment of employees and planning to meet employee needs. Created, organized and provided an ongoing continuing education program.

NURSING EDUCATION INSTRUCTOR
Memorial Hospital, New York, NY 1978-1979

Responsibilities included development and coordination of and participation in orientation programs for nurses, unit clerks, nurses aides, contract agency nurses, affiliating RN and LPN students, and nurse internship program for graduate nurses. Involved in planning, providing and evaluating staff development programs for medical, surgical, progressive care units, and ancillary departments. Member various hospital committees including chairperson Patient Education Committee, Policies and Procedures, Safety, Library, Nursing/Pharmacy Advisory, and Nursing Management.

ASSISTANT HEAD NURSE, SURGICAL UNIT
Harry F. Truman Memorial Hospital, Dover Beach, Maryland, 1975-1978

Responsibilities included: patient care; preceptor for new orientees; assisted with staff evaluation; member of various committees including Patient Education, Nursing Audit and Nurse Recruitment and Retention.

STAFF NURSE, CHARGE NURSE, SENIOR STAFF NURSE/ACTING HEAD NURSE
Fracture Clinic, New York Hospital, NY, 1973-1975

Responsibilities included: coordination and hands-on participation in the care of outpatients and inpatients, supervision of one RN, three plaster technicians, one X- ray technician and two clerks.

Education

Bachelor of Science, Nursing, New York University, New York, NY
Nursing - Accelerated Program

Bachelor of Arts, Washington University, St. Louis, MO
Sociology

Mobile Emergency Medical Technician Course (Paramedic)
Max C. Starkloff Memorial Hospital, St. Louis, MO

Certifications

American Heart Association, New York Affiliate, Instructor Trainer
American Heart Association, Missouri Affiliate, CPR Instructor

Consultant Experience

CPR Instructor Trainer-rewrote syllabi for Basic Life Support courses for Memorial Hospital Training Center 1988 to present

Health Care Education 1986 to present
Created and presented seminars on Home Medical Emergencies to religious organizations, day care providers and civic organizations.

Professional Affiliations

Present: American Nurse's Association
American Heart Association

Past State Nurse's Association - Member of the Council on Continuing Education Review Team 1984 to 1988
New York Inservice Exchange
Dover Beach Heart Association - Speakers Bureau
Member of Dover Beach, Volunteer Ambulance Corps
National Association of Emergency Medical Technicians - Founding Member

GRACE F. ROSS

3985 West Robin Lane
Grand Blanc, Michigan 49116
Home: (313) 352-5748

Results-oriented human resources professional with key accomplishments in management, training, program development and fiscal administration. Experience includes increasing responsibility in both education and corporate environments.

MAJOR ACCOMPLISHMENTS

Management

Directed corporate training function for a multi-bank holding company and managed daily department operations.

Recommended and implemented the reorganization of a corporate training department to more effectively utilize staff and resources.

Streamlined scheduling procedures to eliminate duplication of effort and reduce travel for eleven professionals.

Improved department staffing through effective interviewing, hiring, training and development, and salary administration of exempt and non-exempt staff.

Training

Developed, wrote and implemented supervisory skills modules for adult learners.

Wrote procedures manuals for job-specific skills and trained adults in the appropriate operations.

Taught an American Institute of Banking class for Lake Michigan Community College.

Conducted seminars for a professional women's organizations.

Program Development

Improved a corporate training curriculum by including courses in supervision, communication skills, professional development and personnel issues (EEO/Affirmative Action) based on a needs assessment instrument.

Participated in the development of a bank's retail marketing program in conjunction with a marketing agency.

Directed the activities of a pilot project to improve sales training.

Fiscal Administration

Established and monitored a $1 million training department budget for fiscal 1990.

Contracted with over forty regional banks to increase their level of support for training.

Reduced delinquency rate in an installment loan portfolio through effective customer problem-solving techniques.

PROFESSIONAL HISTORY

Corporation Training Manager
Empire National Bank, Southfield, MI 1986 - Present

Retail Banking Officer
Citizens Trust and Savings Bank, Flint, MI 1984 - 1986

Responsibilities in human resources, branch administration and consumer lending.

Loan Officer
American National Bank, Ann Arbor, MI 1981 - 1984

* Teaching career spanned seven years in Oregon and eight years in the Oakland County, Michigan school systems.

PROFESSIONAL ASSOCIATIONS

American Society for Training and Development

American Institute of Banking

Financial Women International

EDUCATION

M.S. University of Oregon, Eugene, Oregon
Education

B.S. Oregon State University, Corvallis Oregon
Elementary Education

A.A. Lane County Junior College, Eugene, Oregon
Liberal Arts

HOBBIES

Golf, Boating and Gardening

FRANCIS M. PETERS

315 Henry St. Oxford, Michigan 48938 (313) 465-3904

Employment History:

Michigan Manufacturing, Inc., 2899 Green Avenue, Detroit, MI 48976
1977 - Present

PERSONNEL MANAGER

Administered all employee/labor relations programs for a work force of 340 employees engaged in the design, production and distribution of automotive accessories. Provided counseling and advice to employees having personal and/or work-related problems. Developed innovative programs to address a variety of personnel issues including: absenteeism, safety, team-building, quality control and career development. Administered and interpreted labor agreement to insure consistent application. Advised supervisory and management staff on application of contract provisions.

Coordinated and participated in the formal grievance procedure, up through and including arbitration. Administered employee benefit program at the plant level.

Administered all corporate, division, and plant personnel policies and procedures. Counseled and advised supervisory/management staff on policy/procedure application.

Coordinated plant safety and security programs.

Directed compliance with local, State, and Federal laws, EEOC, Worker's Compensation, OSHA, DOL, Wage and Hour--represented plant at agency hearings and proceedings.

Developed and coordinated management training activities; assisted in the administration of career development plans. Involved in recruiting both hourly and salaried personnel.

SUPERVISOR OF INDUSTRIAL RELATIONS

Responsible for manpower coordination and planning. Recruited, interviewed and hired hourly employees. Actively participated in all contract negotiations. Established and maintained viable relations with insurance carriers and state and federal agencies.

* Conducted multi-plant Safety Audits.
* Conducted loss-prevention surveys.
* Received award from Michigan Department of Labor for significant reduction of lost-time hours.

SUPERVISOR OF ASSEMBLY, PACKAGING AND WAREHOUSE OPERATIONS

Directed a work-force of over 70. Assisted in doubling production while reducing personnel by over 35%.

* Cited by management for consistently meeting or exceeding production goals.

Education:

Wayne State University Detroit, MI	Bachelor of Science Industrial Relations
Detroit Business College Detroit, MI	Certificate Business Administration

Memberships:

Elks Club - Current Treasurer
Member of Maple Hill School Board - 1988-1990

CARLA J. DAVIS

148 West 15 Mile
Madison, Indiana 49071
(413) 567-9058

PROFESSIONAL EXPERIENCE:

Mitchell Department Store Company
Over fifteen years of diversified Human Resource management experience in the retailing industry.

HUMAN RESOURCE ADMINISTRATION
Responsible for all human resource related activities at store level to include: staffing, compensation, benefits administration, employee relations and affirmative action, for up to 500 employees.

Saved the company $9,000 through consistent follow-through with the Targeted Jobs Tax Credit Program.

Increased sales at store level by:
-reducing turnover
-adding to and balancing existing sales staff
-cross-training personnel clericals in the screening, interviewing, hiring and placement process
-increasing minimum starting wage for sales consultants

Effectively recruited and maintained qualified minority candidates to meet Affirmative Action goals. Accomplished 100% achievement of goals in both management and hourly categories.

Improved employee productivity and company sales by successfully implementing new bonus incentive program.

Reduced absence pay expense by $76,000 by training front-line supervisors to effectively monitor, counsel and consistently discipline employees.

Conducted E.E.O. workshops. Maintained, on a daily basis, the requirements for E.E.O. according to the legal guidelines for:
-interviewing
-hiring
-placement
-discipline
-termination

Reduced Unemployment Compensation costs through effective, consistent discipline and documentation of employee performance problem issues. Savings to the company ranged from $2,000 - $4,000 per case.

EDUCATION:

Bachelor of Arts, graduated with honors
Major: Business Administration
Indiana University,
Bloomington, Indiana

References Available Upon Request

PAT G. GAINES

3268 Courtland Drive
Lathrup Village, Michigan 48298
(313) 736-0394

PROFESSIONAL HISTORY

T & T, Division of Maldo Industries — 1984-Present
18273 West Grand Boulevard
Detroit, MI 48293

MANAGER OF EMPLOYMENT, TRAINING AND DEVELOPMENT
Human resource generalist functions for a salaried group of 350 research and corporate headquarters employees.

Duties and Responsibilities:
* Wage and Salary Administration
* Executive and Technical Staffing
* Design and Implementation of Human Resource Development Programs
* Health Benefit Claims Administration
* Security and Safety Functions

Accomplishments:
Reduced cost of self-insured employee benefit program. Researched the current third party administration agreements, and eliminated the need for reserves. Developed an internal claims processing department and reduced life insurance claims by 65%.
Developed and implemented company training needs assessment and designed training action plans accordingly.
Standardized procedures for hiring both direct and contract employees.

X-MALL-O Corporation — 1971-1984
2843 Executive Blvd.
Farmington Hills, MI 48338

MANAGER OF PERSONNEL AND INDUSTRIAL RELATIONS
Managed all functions of the Personnel and Labor Relations Department for 250 salaried and 340 hourly employees.

Duties and Responsibilities:
* Workers Compensation Administration
* Fringe Benefit Administration
* Executive and Technical Recruiting
* Affirmative Action Plan
* Grievance Administration

Accomplishments:
Advised management during contraction and amalgamation of units.
Reduced both hourly turnover and accident rate to zero, resulting in significant savings.
Trained supervision resulting in reduced grievance activity with increased production.
Represented company to EEOC, MESC, and arbitration -- all with favorable results.

EDUCATION

Bachelor of Business Administration
University of Indiana
Major: Personnel and Production Management
Minor: Economics and General Business

University of Detroit
School of Law
Classes in contracts, procedures, torts crimes, property and business law.

FREDERICK H. BALDACZAR

1122 Gumption Drive
Centerline, Michigan 48097
(313) 735-8765

Extensive manufacturing management experience in variety of settings. Effective, progressive negotiator with proven results.

Employment History:

Michigan Manufacturing, 2899 Green Avenue, Detroit, MI 48976
1975 - Present.

SUPERVISOR OF INDUSTRIAL RELATIONS

Responsible for manpower coordination and planning. Recruited, interviewed and hired hourly employees.
Actively participated in all contract negotiations.
Established and maintained viable relations with insurance carriers and state and federal agencies.

* Established Health & Safety Program.

* Conducted multi-plant Safety Audits.

* Conducted loss-prevention surveys.

* Received award from Michigan Department of Labor for significant reduction of lost-time hours.

SUPERVISOR OF ASSEMBLY, PACKAGING AND WAREHOUSE OPERATIONS

Directed a workforce of over 40. Assisted in doubling production while reducing personnel by over 35%.

* Cited by management for consistently meeting or exceeding production goals.

* Twice awarded for safety record.

Education:

Wayne State University Detroit, MI	Worker's Compensation, Unemployment Compensation, Industrial Safety Management.
Detroit Business College Detroit, MI	Certificate Business Administration

Memberships:

Elks Club - Current Treasurer
Member of Maple Hill School Board - 1988-1990

Hobbies:

Tennis, Golf, and Sailing

LEONE H. CARUTHERS

8377 East Meadowlake Road
Minneapolis, Minnesota 65748
(612) 453-7643

Professional Experience

HUMAN RESOURCES COORDINATOR/ADMINISTRATION
Carter Corporation, St Paul, Minnesota 1986-present
Reporting to the Vice President, Human Resources

Provide Corporate Human Resources information on a company-wide basis to managers, employees and outside consultants/clients by dealing with the response and/or referral of department inquiries, ensuring a confidential, high level of service.

Manage the Expatriate Program, currently 15 employees, including both Canadian expatriates and Third Country Nationals. Responsibilities include preparation of $1.5 million budget, all payroll functions, work permit applications, employee orientations, income tax payments, implementation of cost of living and exchange rates. Notify host countries of any changes in compensation to be paid overseas and invoice on a monthly basis.

Manage and perform all payroll functions for 25 senior executives. Responsibilities include calculation and payment of stock options, benefits, salary and bonus budgets, and bonus eligibility.

Over an 15-month period, coordinated the implementation of a national reorganization of the Canadian company, including the downsizing of 40 employees, in consultation with the President, Vice President and external consultants.

Facilitate training classes for the Carter Quality Process.

Hire, train and manage clerical support for the Human Resources department.

* Proficient in: Lotus, Harvard Graphics, Display Write, and WordPerfect.

ASSISTANT TO VICE PRESIDENT
HuTech Temporary Services, Minneapolis, Minnesota 1975-1983

Reporting to the Vice President, Operations; responsibilities included: all administrative functions, drafted correspondence, appointments scheduling, and arrangement of meetings and management seminars. Organized and managed annual receptions for approximately 1000 clients. Prepared monthly reports documenting call activity, weekly sales, profit and loss.

Previous Employment: administrative assistant, secretary, receptionist

Award

Human Resource Professional (HRP) designation awarded by Personnel Association of Minnesota, 1990

Education

Certificate in Personnel Administration (CPM)

Completed training programs in the following areas:
Payroll Administration Basic Management Principles
Word Processing Training Presentation Skills

Certificate Program, Applied Sciences, Sherwood College, St. Paul, MN

JANE L. BAKER

2427 State Street
Lawrence, Michigan 40937
(313) 555-9340

EMPLOYMENT HISTORY:

Lexington Industries, Inc.
I-75 & Plain Road
Pottstown, MI 43807
1978 - 1990

BENEFITS ANALYST
Processed medical claims at the Pottstown Division of Lexington Industries. Reviewed documents submitted by employees and by providers and calculated benefits payments. Duties also included administration of complex coordination of benefits and handling questions on the employee hotline. Operate CRT, IBM PC and electronic calculator.

* Trained new Benefits Analysts.
* Received award for excellent job performance.
* Participated in conversion of system from manual to computer.

Other positions at Lexington:

clerk-typist, stenographer, receptionist.

SNR Industries
Pottstown, MI 43807
1971 - 1978

CLERK-TYPIST
Performed general clerical duties in Personnel Department of foundry.

EDUCATION:

Pottstown Business Institute
Completed Secretarial Skills program.

Pottstown High School
Graduated with honors
Business curriculum

HOBBIES & INTERESTS:

Secretary of Sunday school
Sailing, bowling and gardening

PERSONAL DATA:

Willing to relocate
Excellent health

References Available Upon Request

GERALDINE J. CASHMAN

21113 Woodland Glen Drive #102
Seabrook, Texas 98671
(713) 234-9209

Training and Development professional with extensive human resource experience in two different markets for one of the country's largest retailers. Specialized in insuring the best value for the training dollar spent by operational needs assessment and tailored programs.

Employment History

1975-Present

The Scott/Marshall Co.
Human Resources Department, Houston, Texas

MANAGER OF TRAINING AND DEVELOPMENT 1987-Present
Promoted to spearhead training and development efforts during intense market expansion. Supervise development of orientation and basic training programs for over 4500 employees in 6 in-store departments, for 50 store locations. Conduct successor planning and manpower forecasting for 135 management level personnel. Recruit store and pharmacy management personnel from campuses and stores. Coordinate staffing for new combo store, including hiring and training over 275 employees.

* Developed and facilitated supervisory skills seminars and train additional facilitators.
* Structured and facilitated "The Complete Manager" management course.
* Responded to EEOC/MDCR charges including settlement of one charge which avoided a $10,000 cost to the company.

ASSISTANT FOR TRAINING AND DEVELOPMENT 1984-1987

Developed and implemented a primary cog of a multi-faceted plan to increase meat sales by 10% and improve reputation. Achieved goal by training facilitators to conduct customer service training for over 350 associates, developing a division-wide schedule, and managing program costs. Supervised management training program, 70 graduates annually. Developed and facilitated "Train the Trainer" sessions for over 125 trainers. Conducted communications training seminars for over 200 store management personnel, in marketing area and corporate level. Budgeted and monitored expenditures of $1.5M in training funds. Revamped the Scott/Marshall Company corporate management training program, as one of six individuals selected from throughout the company for this responsibility.

SENIOR ASSISTANT FOR ADMINISTRATION/LABOR RELATIONS 1982-1984

Administered health care and pension benefits for over 12 company and trusteed plans. Supervise records system for 4500 employees. Insured regulatory compliance of 54 store locations. Supervised management assessment and records transitions of 35 acquired stores. Developed positive employee relations through equitable labor contract administration in 10 stores.

STORE MANAGEMENT 1975-1982
Customer Service Manager, Iron Mountain, CO
$300,000/week sales, over 140 employees

Co-Manager, Boulder, CO
$250,000/week sales, over 120 employees

Previous Employment includes: Department Manager, Clerk, Inventory Clerk, Personnel Clerk, Secretary, Receptionist

Education:

M.A. - Major: Organizational Communication
Western Texas University

B.S. - Texas A & M

Professional Affiliations:

American Society for Training and Development
Society for Human Resources Management

NORTON P. JEFFRIES

3857 Wyandotte Drive
Troy, Michigan 48084
(313) 738-3049

Skilled Attorney with extensive experience in Taxation, Commercial Law, Corporate Law, Estate and Tax Planning, Real Estate and Construction Lien Law.

PROFESSIONAL HISTORY

Lambert, Wilson, Docket and Smythe
Troy, Michigan
1981-Present

ATTORNEY AT LAW
Broad spectrum of corporate and tax law experience with both manufacturing and service corporations. Extensive consultation, planning and research in Tax, Corporation, Real Estate, and Commercial Law, and Litigation, as well as analysis and drafting of related programs and strategies.

* Successfully advised numerous corporations on taxation and legal strategy in business and decision making, resulting in substantial savings and leverage for future investment.

* Continually sought by manufacturing clients for ability to problem solve for increased productivity and profit.

Wayne State University, College of Business
Detroit, MI
1985-Present

INSTRUCTOR
Coordinate legal research for the College of Business. Teach graduate and undergraduate courses in Business and Commercial Law.

* Responsible for research on landmark Michigan class-action lawsuit regarding non-refundable cleaning fees.

Myers Manufacturing Company
Detroit, MI
1976-1981

ASSISTANT TO THE PRESIDENT
Advised top management in tax and legal areas.

* Succeeded in prevailing in major Internal Revenue Service Audits.

* As Management/Employee Relations Liaison, implemented corporate policies, mediated employee grievances, and coordinated all facets of the production process.

EDUCATION

Juris Doctor - The University of Michigan, Ann Arbor, MI
M.B.A. - Wayne State University, Detroit, MI
Bachelor of Arts - Albion College - Phi Beta Kappa

COMMUNITY SERVICE

Legal Counsel for the Troy, Michigan Jaycees

SYLVIA P. CARDOMAN

3894 Peachtree Lane
Hartford, Connecticut 99760
(203) 965-8059

EDUCATION

University of Connecticut School of Law, J.D., 1991, Awarded with Honors
University of South Carolina School of Law, visiting student, 1989 - 1990

University of Chicago School of Social Service Administration, one year toward M.S.W., 1964 - 1965

Bryn Mawr College, A.B. in Anthropology, 1962, awarded cum laude

ADDITIONAL TRAINING

Divorce Mediation training - Erickson Mediation Institute, meets certification standards of Academy of Family Mediators, 1990

HONORS

American Jurisprudence Awards, awarded for receiving the highest grades in Constitutional Law and Criminal Procedure, presented 1991

"Municipal Liability in South Carolina," paper included in the Student/Faculty Collection of the McKusick Law Library at the University of South Carolina, 1990

Dean's List, University of South Carolina School of Law, 1990

Certificate of Recognition, American Bar Association, Section of Urban, State and Local Government Law, 1990

PROFESSIONAL EXPERIENCE

LEGAL INTERN
University of Connecticut Civil Clinic, summer 1988
Conducted legal research and client counseling.

TEACHER
Art Museum Nursery School, Hartford, CT, 1985 - 1986
Designed curriculum and supervised a group of preschool children.

ASSISTANT DIRECTOR
YMCA Day Care Center, Hartford CT, 1984 - 1985
Assisted with program coordination, designed and evaluated curriculum and worked with preschool children.

References and Writing Samples Available Upon Request

GERRY M. HEMMINGS

3746 Dawn Avenue
Detroit, Michigan 48764
(313) 829-0192

Self-motivated manager with strong technical background; knowledgeable in all aspects of Production Management. Skilled in diverse areas: budgeting, team-building, training and quality control.

WORK EXPERIENCE:

XYZ, Inc., 3543 Maple Ave., Detroit, MI
(Available due to plant closing.)

PRODUCTION MANAGER 1985 - Present

Managed all aspects of several departments involved in packaging and processing of various personal care products. Directed production activities in support of sales needs. Monitored and controlled departmental spending and supply usage within budgeted allowances. Responsible for monitoring overall operation of department in safe manner, consistent with both internal and governmental requirements.

Prepared and reviewed departmental operating budgets ($1.5M) and coordinated labor/capital estimates for new or revised products.

Responsible for personnel-related activities; staffing, training, and discipline. Directly supervised 9 departmental employees.

PROCESSING SUPERVISOR 1983 - 1985

Coordinated and supervised various departments to insure availability of raw material in support of production schedule. Responsible for maintaining productivity of workers to control direct labor cost of compounding formulation within budgeted constraints. Controlled overhead expenses with special emphasis on indirect labor and supplies.

Initiated and conducted several safety awareness programs, resulting in the lowest accident/employee ratio for three years in a row.

Worked cooperatively with Material Control Department to order and receive bulk chemicals as needed.

Maintained alcohol usage records in accordance with Federal regulations.

- Promoted to Production Manager after only two years
- Excellent performance evaluations

ABC International, 463 Avon Lane, Brighton, MI

SUPERVISOR 1979 - 1983

Responsible for the supervision and coordination of 55 production workers engaged in the manufacturing of high-quality automobile tires. Handled all hiring, training, scheduling and grievances.

Utilized knowledge of layout, product design, machine functions and output potential.

Additional duties included the requisition of material required to meet production goals; monitored quality and maintained production records.

EDUCATION: Bachelor of Science - Chemical Engineering
Wayne State University
Detroit, Michigan

AVERY A. EDBERG, P.E.

14151 Upland Avenue
Baltimore, Maryland 12431
Residence (418) 353-3651
Business (418) 336-3045

In-depth business and project management experience in all phases of manufacturing with American Soap. Expertise in business unit management, total quality management, and productivity/cost reduction improvements. Master of Science degree, M.I.T.

EXPERIENCE

American Soap Manufacturing Co. 1979 - present
(Available due to plant closing.)

Manager, Shutdown Operation (1991 - present)
Direct all site clearance activities for bar soap operations, including equipment removal, transfer or disposal; environmental remediation; manage $60 million shutdown budget. Supervise 35 technicians and 10 managers.

Achievements

- Will complete project on time and 15% under budget.
- Decreased scrapping costs by $1.5 million.
- Zero safety incidents during shutdown to date.

Technology Manager - Clear Bar Soap (1988 - 1991)
Production and maintenance responsibility for national Clear Bar Soap manufacturing facility. Directed mechanical organization to a team based high performance work system; implemented reliability program using total quality principles. Managed annual capital and expense budget of $7 million for business with $45 million in annual sales; developed technician owned preventive maintenance/spare parts program; trained and coached 8 technicians and 4 managers.

Achievements

- Increased efficiencies by 22%, saving $525,000 per year through team skill training.
- Reduced craft classifications from five to one while maintaining work group skill capabilities.
- Reduced sick time from 11% to 3% through initiation of new absenteeism program.
- Implemented special assignment and functional leadership roles to move management duties to technicians.

Project Manager - Plant Engineering (1984 - 1988)
Responsible for projects in excess of $18 million; developed cost-effective capital funding and execution strategies; provided complete project leadership.

Achievements

- Project Manager - $8.5 million project to design and construct national facilities to produce new synthetic detergent product. First project executed using Total Quality principles. Delivered on time and under budget.

- Project Manager - New $2.5 million production facilities for national Lemon Cleanser.

Business Manager - Clear Soap (1982 - 1984)
Responsible for national production of Clear Bar Soap including: budget development and control; cost and productivity improvement; recruiting, training, and development of personnel; union relations and safety. Supervised 45 technicians and 5 managers.

Achievements

- Directed national consolidation of Clear to Eagle Island location.

- Increased volume 55% by growing to 24-hour operation. Maintained all quality goals.

- Decreased unit cost by 15% through volume and productivity improvements.

Supervisor - Soap Process Department (1979 - 1982)
First-line supervision of all operational and mechanical functions for fat and oil processing to make base soap for the bar soap departments.

Story & Westman Engineering Corporation 1974 - 1979
Baltimore and Washington

Progressed through assignments of increasing responsibility on nuclear power plant projects. Areas included: quality assurance; structural specification/design review; coordination of project administration; budget/cost/schedule review; and client relations.

EDUCATION

Massachusetts Institute of Technology
M.S. Degree in Project Management, 1978
Minor in Business Administration, Sloan School of Management.

Georgetown University
B.S Degree in Engineering

Maryland State Professional Engineer License No. 045608

KEITH T. DECKER

5874 South Glendale Drive
Farmington Hills, Michigan 49384
(313) 395-2497

Experienced manager with over twenty-five years of background in Stamping Operations from Plant floor to Engineering Management. A self motivated, team player successful in directing multi-million dollar programs and Plant expansions while maintaining production of Stampings and Assemblies for Car Assembly Operations.

Directed Plant skilled work force, engineering areas and future planning of major programs at staff level. Major emphasis on Process Design and Implement Systems to achieve world class stamping at competitive costs.

SELECTED ACCOMPLISHMENTS

HUDSON MOTOR CORPORATION
Manager, Methods, Technology & Administration 1987 - Present
Developed new technologies in die process and sub-assembly processes to reduce investment and piece cost, as well as related cost reduction objectives.

Developed methods and practices to reduce new model lead time from 4 years to 3 1/4 years.

Revamped group stamping production engineering organization for increased effectivity as company down- sized.

Stamping Productivity Study 1987
Traveled world wide reviewing competitive processes and suppliers for potential cost reduction and implementation.

Project Management 1986 - 1987
Responsibility included coordinating teams to effect major launches that met corporate objectives.

Task Force Leader 1976 - 1986
Organized productivity teams at (2) locations.

Conducted (9) OSHA facility inspections as member of OSHA audit team.

Manufacturing Manager 1983 - 1985
Completed $15 million expansion and renovation program.

Installed and launched (15) new sub assemblies.

Led all 425 skilled trades people on new model launch and production support activities, while maintaining production requirements.

Tool & Die 1982 - 1983
As Production Engineering Manager, led tool and die operations on new model construction, tryout and launch. Resulted in quality on time launch within cost objectives while supporting production operations.

Facilities Management 1981 -1982
Under severe financial and time constraints created and compiled facility improvement projects as follows:

- Building and equipment maintenance
- Automation installations
- New press installations
- Sub assembly installations
- Disaster preparedness plan

Stamping Tool/Machine Process & Design 1978 - 1981
Supervised process and design in stamping plant, including tool follow up.

Maintenance Superintendent responsible for tools maintenance activity.

Die Shop Superintendent responsible for department activities (3) shifts.

OTHER PROFESSIONAL EXPERIENCE

Die Maker * Die Process * Die & Tool Design Supervisor * Maintenance Supervisor * Project Leader * Facilities Manager * Program Planning * Lead Time Reduction

EDUCATION

Lane University - Certificate, Engineering Administration

ADDITIONAL TRAINING

Lansing College of Applied Science - Certificate, Tool and Machine Design
Hudson Motor Tool & Die Apprentice - Graduate Die Maker
Dale Carnegie Leadership Graduate
Time Management - Franklin Institute
Technical Writing

HOBBIES

Outdoor activities including boating, fishing and golf

PROFESSIONAL ASSOCIATIONS

Engineering Society of Detroit
Hudson Motor OSHA Task Force

WILSON G. DUNLAP

3897 North County Road 56
Indianapolis, Indiana 48901
(318) 425-4244

PROFESSIONAL WORK HISTORY

American Star Corporation, Indianapolis , IN
1975 - 1991 (Available due to reduction of management work force.)

Machining Supervisor - Direct supervision of 25 production workers operating 30 machines. Responsible for machining rough stock aluminum into finished clutch retainers and pistons for assembling into state-of-the-art automobile transmissions. Work closely with Engineering and Maintenance Departments in problem-solving techniques since this is a new machinery line.

* Developed effective management skills, included working in union environment, motivating workers, and disciplining. Good working relationship with union officials, workers, fellow supervisors, and upper management.

* Consistently met production goals. Never responsible for stopping assembly build line valued at $13,000 per minute.

* Consistently met cost reduction/cost avoidance goals. Reduced scrap, improved housekeeping, decreased assembly calls, eliminated assembly "send homes."

* Implemented Statistical Process Control (SPC) which generated $37,000 savings per week.

Machine Operator - Joined American Star immediately following graduation from college. Operated grinders, industrial drills, straighteners and broaches.

* Promoted to supervisory position of newly-developing department after only four months operator experience.

OTHER WORK HISTORY

Dairy Queen, Inc. - Kokomo, IN
Assistant Manager, summers

* Worked 15-20 hours per week while attending Ohio State University as full-time student.

EDUCATION

Ohio State University
Columbus, Ohio

B.S. Degree
Management

PERSONAL

Married, excellent health. Willing to travel. Willing to relocate.

CHARLES F. MALONEY

38953 Maplevale Road
Ft. Lewis, Tennessee 97812
(901) 733-3903

EXPERIENCE:

Star Electronics, United Motors, Ft. Lewis, Tennessee — 1984 to Present

PRODUCTION FOREMAN
FACILITIES ENGINEER
MAINTENANCE ANALYST
ENGINEERING CLERK
MAINTENANCE STAFF

In six years promoted from entry level maintenance position into management due to hard work, dedication, good people skills and continuing education.

Provided equipment and facilities planning including design, layout, modification and maintenance of 15,000 square foot highly automated electronics manufacturing and engineering facility. Controlled contracted plant modification and maintenance services. Prepared cost estimates, designed and directed installation of plant HVAC systems. Responsible for plant telecommunication system including all voice and data networking. Supervised employees in manufacture of electronic printed circuit boards. Scheduled all assembly activities. Responsible for production and quality to meet strict military specifications.

* Reduced production line solder defects by 50%
* Established employee cross-training program
* Developed wave soldering parameters
* Initiated plant wide preventative maintenance program
* Worked full time while completing degree in Mechanical Engineering

Towne Club Bottling Co., Ft. Lewis, TN — 1976 to 1984

MAINTENANCE REPAIRMAN
Responsible for mechanical and electrical repair of high-speed production equipment.

Maintained ammonia and freon refrigeration systems, automatic lubricators, conveyors, air line system, compressors etc. Scheduled and implemented preventative maintenance program.

Parker Supply, Ft. Lewis, TN — 1974 to 1976

WAREHOUSE SUPERVISOR
Responsible for warehousing, shipping, and receiving for a wholesale plumbing supply firm.

EDUCATION:

Ft. Lewis College, Ft. Lewis, TN
B.S. - Mechanical Engineering

Calhoun Community College, Decatur, Alabama
Mathematics and Pre-Engineering

TECHNICAL TRAINING:

Statistical Process Control
Telecommunications, South Central Bell
Electronics, Drake Technical College
Industrial Electricity/Electronics, Drake Technical College

MICHAEL C. MARVELOUS

3044 White Birch Drive
Old City, Missouri 35980
(329) 333-8907

EMPLOYMENT HISTORY

STRUCTURAL SUPERINTENDENT - Steel Fabrications Co., Old City, MO 35980
1984 - Present, (available due to plant closing).

Overall production and budget responsibility for all structural fabrication and welding departments. Directed workforce of 10 foremen and 230 hourly employees. Prior to structural assignment, served as Storeroom Superintendent and Second Shift Plant Superintendent.

* Brought the department on schedule on all large weldments.
* Rebuilt the department's group of supervisors into an excellent first-line team.
* Implemented and coordinated a Quality Circle Employee Involvement Program.
* Increased Storeroom inventory accuracy by 30%.

GENERAL FOREMAN - Marsh Welding Corp, Milwaukee, WI 53212
1979 - 1984

Began career as an Industrial Engineer developing data and methods for standard hour incentives with significant success in reducing production costs. Accepted a position as Foreman in the fabrication and welding departments.

Promoted to General Foreman, supervised 5 foremen and 15 hourly employees; continued to solve production problems in addition to planning and scheduling.

EDUCATION

Associate in Applied Science
Michigan School of Engineering, Lansing, Michigan

Industrial Engineering Technology
Credits toward degree in Industrial Management and Mechanical Engineering.

MILITARY

United States Army - Stateside and Abroad
Sergeant - Served as an infantry squad leader, Second Infantry Division
Honorable Discharge

PERSONAL

Excellent Health * Willing to Travel and Relocate

HAROLD P. ANDREWS

5348 Oak Street
Kenosha, Wisconsin 53142
(414) 654-9039

EMPLOYMENT HISTORY:

Wisconsin Manufacturing, Kenosha, Wisconsin 53142
1976 - Present, (available due to plant closing.)

QUALITY CONTROL LAB TECHNICIAN
Participated in activities concerned with development, application, and maintenance of quality standards.

Responsible for adhering to specific methods and procedures for inspection and testing of raw materials and stamped parts at various stages of production. Familiar with computerized quality control systems.

Participated in trouble-shooting and made recommendations to Quality Control Supervisor.

Other duties included: recordkeeping, written reports and file maintenance.

- Commended for accuracy and attention to detail.
- Excellent work record.
- Awarded for perfect attendance three years in a row.

MILITARY HISTORY:

U.S. Army, Honorable Discharge

EDUCATION:

Taylor High School	Graduated, with honors
Oakdale, WI	Industrial Arts

HOBBIES AND INTERESTS:

Hunting, Golf, Fishing
Remodeled house built in 1910.

VOLUNTEER WORK:

Boy Scout Leader - 10 years
Awarded for outstanding service by Boy Scouts of America

PERSONAL DATA:

Excellent health * Married * Three children
Willing to travel and/or relocate

References available upon request

EDWIN W. EDWARDS

6726 Strong Drive
Indianapolis, Indiana 46217
(313) 487-3963

In-depth ferrous foundry metallurgy, process control, and quality control background. Considerable experience in casting design development, material and process standards development, materials cost control, foundry process control (melting, molding, and core making), laboratory management, quality problem solution, and quality control.

EXPERIENCE

POWELL CORPORATION
Michigan and Indiana

Through 1990

QUALITY CONTROL - RESIDENT ENGINEERING MANAGER

- Responsible for the product quality control at the Indianapolis Foundry producing gray iron cylinder block castings.
- Managed Quality Engineering, Inspection, customer contact, Process Control, Product Engineering, and Materials Laboratory functions.

QUALITY CONTROL - RESIDENT ENGINEERING MANAGER

- Responsible for Product Engineering, Quality Control, Materials laboratory, Inspection, and Process Control at the Bronx Avenue Foundry, Detroit, Michigan. Automotive foundry employing over 3,000 which produced gray iron cylinder blocks, cylinder heads, and camshafts as well as ductile iron crank shafts.
- Coordinated materials and product design changes with corporate engineering and customer plants.
- Identified, defined, and corrected quality problems.

RESIDENT ENGINEERING MANAGER

- Responsible for Materials Engineering, Material Laboratory, Process Control, direct material cost reduction, and Product Design Engineering.
- Reduced productive material cost of plant by over $1 million annually.

PLANT METALLURGIST

- Supervised 35 salaried employees staffing the Process Control, Receiving Inspection, and Materials Laboratory.

LABORATORY SUPERVISOR AND MATERIALS DEVELOPMENT ENGINEER

- Managed Heat Treatment Laboratory and developed ferrous castings for automotive application at the corporation's Central Engineering Division in Detroit, Michigan.

MILITARY SERVICE

- United States Army, Captain, 34th Tank Battalion, Tank Platoon Leader and Operations Officer. Honorable Discharge.

EDUCATION

- B.S., Metallurgical Engineering, Michigan State University. Graduated with Honors.
  Member Tau Beta Pi, National Engineering Honor Society.

STANLEY J. BOWDEN

13423 14 Mile Road
White Cloud, Michigan 49839
(517) 738-3894

EMPLOYMENT HISTORY:

American Automotive Parts, Inc. 1432 Maple Leaf Drive, Farnsworth, MI 49039 June 1973 - Present
(available due to permanent plant closing)

QUALITY CONTROL AND INSPECTION
Employed by a major manufacturer of quality automotive parts and accessories.

Responsible for spot checking products for critical, major or minor defects and conformance to tolerances and specifications as set by Inspection Department.

Maintained sampling reports, monitored rejects and filled out deviation slips when required.

Qualified to read blueprints and use all necessary gauging equipment.

* Excellent attendance and safety record.
* Commended by supervisors for maintaining a clean work area.
* Successfully trained over 35 new employees.

Acme Automotive, RR #31, Grand Rapids, MI
June 1967 - June 1973

GENERAL MACHINE OPERATOR
Member of a skilled team engaged in the production of automotive parts and wiring harnesses.

Responsible for the operation of the following:

* Punch Presses
* Bending Equipment
* Spot Welders
* Fork-lift

Experienced in the following functions:

* Assembly
* Checker
* Inspection
* Material Handling

Consistently exceeded all production quotas.

HOBBIES AND INTERESTS:

Fishing, woodworking and gardening

LUCIUS P. ANGELINO

38951 River Rock Drive
Oakdale, Kentucky 58987
(606) 286-4979

ACCOMPLISHMENTS:

* Built sales in every store by improving store conditions, customer service, and merchandise availability.

* Managed store that was selected for manager training. Over a period of five years, was mentor to numerous trainees entering store management.

* Selected to manage two new stores and three completely remodeled stores.

* Conducted weekly store board meetings with Co-Manager and Department Heads to review merchandising, sales planning and operating problems; kept all employees informed in order to ensure maximum teamwork.

* Regularly observed competitors' activities and took appropriate action, through headquarters, to meet the competition through special promotions.

* Developed and maintained a reputation as a firm but fair manager with high expectations of the store staff.

BUSINESS EXPERIENCE:

STORE MANAGER, HOLIDAY MARKET COMPANY 1976-1990
Full responsibility for the profitable operation of high sales volume supermarket in Louisville, KY. Responsibilities included:

Staffing: Interviewed, hired, and trained employees. Administered union contracts (Retail Clerks and Meat Cutters). Evaluated employee job performance and took appropriate action to achieve standards. Supervised scheduling. 100+ employees.

Merchandising: Supervised all ordering of both perishable and non-perishable merchandise from Holiday warehouses and outside vendors. Ensured freshness and quality of product through vigorous monitoring of ordering and rotation. Planned and implemented sales plans, product presentation, and special promotions. Managed merchandise assortments appropriate to store's ethnic mix.

Customer Service: Supervised maintenance of inventory levels to assure availability of product. Supervised check-out function to assure prompt, accurate and courteous customer service. Personally handled customer questions and/or complaints to assure complete satisfaction.

Additional Responsibilities include the maintenance of high standards for sanitation, compliance with government regulations and Holiday policies, and managing the store's physical plant.

PREVIOUS WORK HISTORY: Produce Manager, Meat Manager, various other positions in the retail industry.

EDUCATION: Oakridge Community College
Associates Degree - Management

ELIZABETH W. MURPHY

324 West 19th Street
Kenilworth, New Hampshire 08854
(201) 463-9055

Highly motivated and knowledgeable sales professional with strong technical background and proven management skills.

Work Experience

ASSISTANT MANAGER
Techland Computer Center, Kenilworth, NH — 1985 to present
(A franchise of Micro-Point Dealership)

Responsible for sales of hardware and software for IBM, NEC, Epson, Hewlett Packard, ALR, AST, Compaq and networking products. Developed corporate accounts, maintained customer relations, closed statements. Managed purchasing, retail merchandising, software demonstrations.

MATERIALS MANAGEMENT
Micro Point Computers, Greenbrook, NH — 1983-1985
(An affiliate of Micro-Point of America, New York, NY)

Responsible for purchasing, processing computer and product orders, client billing, accounts payable, accounts receivable, and inventory management. Assisted sales department with product pricing, availability of products and product information.

SALES MANAGER/EXECUTIVE BUYER
Billie Jeanne Collections Inc., Queens, NY — 1982-1983

Responsible for management and supervision of staff and trained new sales employees for both locations. Supervised shipping/receiving departments, kept daily sales records, kept daily sales panafax, and organized store merchandising and display, Involved with all aspects of high fashion business including print advertising and public relations.

ASSISTANT BUYER
Software Gallery, Concord, NH — 1980-1982

Responsible for purchasing software, order processing and assisting sales department with pricing and product availability information.

ASSISTANT MANAGER
Bobby's Jewelry, New York, NY — 1978-1980

Responsibilities included sales, customer relations, display-merchandising and banking.

SALES ASSOCIATE
House of Pierre, Ltd., New York, NY — 1976-1978

Responsibilities included sales of high fashion women's clothing.

Education

Bachelor of Arts, New York University
Majored in Communications

Hobbies

Tennis, Raquetball, Swimming

DAVID L. DUPUIS

38927 Waddington
Evanston, Illinois 28821
(323) 335-2894

In-depth management experience in restaurant operations, hospital food service, hotel food service, and hotel/motel operations. Demonstrated abilities in the following areas: supervision, training, operations, public relations, and customer service.

WORK HISTORY:

GENERAL MANAGER
PG & L Pizza Pie, Inc.,
Evanston, IL
1978 - 1991

Full P&L responsibility in 2 fine dining and 4 fast-food establishments. Managed workforce of over 200 employees ... hired, trained, supervised, conducted union relations. Controlled cost of food, liquor, and labor. Maintained sanitation. Controlled theft. Prepared budgets and maintained business records.

- Increased weekly volume of primary location from $11,500 to $16,500 despite a depressed economy within the city.
- Increased weekly sales volume of Daisy's 20% in two-month period.
- Developed and implemented a control system in hotel dining room which reduced theft of wine from 100 bottles per month to zero.
- Designed innovative scheduling system to decrease turnover and stabilize workforce.

Prior to 1978: Managed food service in 350-bed hospital. Managed family-owned 202-room hotel. Managed three motor inns.

EDUCATION:

Bachelor of Science, Communications Major
Michigan State University
East Lansing, Michigan

HOBBIES AND INTERESTS:

Tennis, Swimming and Golf

VOLUNTEER WORK:

American Cancer Society - Local Unit Treasurer

PERSONAL DATA:

Willing to travel/and or relocate

JULES S. AUGUSTINE

9485 Hollings Drive, N.W.
Sterling Heights, Michigan 48397
(313) 734-0394

Experienced SALES DIRECTOR; in-depth knowledge of metal cutting tool industry. Special strength in hiring and training effective regional sales managers and developing strong distributor network.

DIRECTOR OF SALES 1977 - Present
ABC Technical, Inc., Madison Heights, MI (Available due to corporate merger.)
Established territories. Recruited and trained sales force. Provided continuous advice and support to regional managers, distributors and distributor salespeople concerning engineering data and sales techniques. Analyzed market to determine customer needs, volume potential, and pricing strategy. Developed and implemented marketing plans, including advertising and trade show exhibits. Directed the development of sales and technical manuals. Responsible for realistic forecasting and sales expense. Personally participated in sales calls to major and potential accounts.

* Major Accounts. Developed Hi-Drive and Platz & Wiley to #1 and #2 accounts, both at a high of $1,750,000 annually. Also developed into major accounts: Columbus Milacon, Drits & Sabbon, Tennessee Instruments, and American UnitTech.

* New Product Introduction. Examples of major impact new product development:

-Boring Bar Set. In 1985, when big ticket items were hard to sell, developed bar set priced at $495. Developed special sales incentives for distributors and salespeople. Seven months sales volume: $400,000.

-Twin Bore. In 1984, conducted total market study of product manufactured by our company in England, resulting in decision to market in U.S. Worked with Engineering, R & D, and other staff to prepare tech manuals, advertising, pricing, and stock prior to announcement. Annual sales: $600,000.

-Pre-set Machines & T.M.S. - 1983. Successfully introduced a complete line of pre-sets developed by our English company. Sold 88 machines in one year, compared with previous sales of 15-20.

* Development of Distributors. Completely revamped distributor network, eliminating non-productive distributors and developing winners. Trained five Regional Managers who, with further training and assistance, became successful distributors or managers of our major distributors. Trained new distributor salespeople by bringing them into the plant and orienting them to all facets of the business. Personally worked in the field with all distributors.

REGIONAL SALES MANAGER
Carmel Division of Aerotech International, Madison Heights, Michigan. 1975-1977

EDUCATION: B.A., Michigan State University

BOBBY SAMUELS

9071 West Wren Road
Cumberland, Tennessee 97365
(901) 765-7638

PROFESSIONAL EXPERIENCE:

MARKETING RESEARCH ANALYST
American Signal, Corp., Aftermarket Division
102 Pucket Ave., Providence, TN 92916
1983 - Present

Provide management with relevant marketing information used in strategic planning, product and new market analysis and problem solving.

Manage various primary and secondary market research studies and data requests from inception through analysis to recommendation.

Function as business/marketing consultant to product management teams for assigned product lines.

Quantify and project various pieces of market information: market size, market share, channel share and vehicle segment for assigned product lines.

Supervised and initiated department's transition to increased use of computer systems. Developed a variety of computer applications used to report on trends within the marketplace.

Promoted from Marketing Assistant to Programmer to Forecast Analyst and finally to Market Research Analyst in only 5 years.

Hardworking employee, with strong analytical skills, enjoys challenge and learning.

EDUCATION:

Master of Business Administration
Marketing emphasis
University of Tennessee
1984 - 1986

Bachelor of Business Administration
Cumberland College
1979 - 1983

MEMBERSHIPS:

Automotive Market Research Council

Forecast Committee 1987 - Present

HOBBIES:

Swimming, Golf, Skiing, and Camping

CAHILL M. MATSON

2849 West Oak Drive
Elmhurst, Indiana 47384
(428) 387-4039

PROFESSIONAL EXPERIENCE

Navco Industries, Inc. 1979-Present
1726 West Jefferson Ave., Elmhurst, Illinois

Regional Sales Manager
Responsible for 35 accounts in 8 Northwestern states. Managed all staff involved in sales and service. Conducted training sessions, annual meetings.

* 28% increase in sales in the first year.
* Reconciled all debts and financial issues.
* Signed up 9 new accounts and brought 6 lost accounts back on board.

Sales & Service Representative
Responsible for 27 accounts in 12 Western states, Mexico, British Columbia and Alberta, Canada. Researched major market areas for new accounts. Provided input from field on competitive products. Introduced new products and maintained regular contact with all accounts. Consistently met or exceeded all sales quotas. Coordinated major national trade shows.

* Increased sales for 8 years consecutively.
* Number one in sales for 8 out of 9 years, while representing smallest market area.
* Represented highest profit margin with least amount of bad debt and least amount of customer turnover.
* Resolved all legal issues with retail customers without further litigation.

Production Plant Manger
Responsible for managing 53 employees involved in the production of OEM Mustang T-roof program for Ford Motor Company.

* Reduced work force by 20% and increased production from 8 vehicles per day to 45.
* Increased quality to meet or exceed Ford's Quality Job 1 standards.

Production Plant Supervisor
Supervised 78 employees involved in the manufacturing and production of the 1979 OEM T-roof program for Ford Motor Company.

* Averted mass employee walk-out, resolving concerns and resuming production in less than 24-hours.
* Personally trained all assembly personnel.
* Reduced man hours per vehicle from 10.2 to 4.6 while maintaining the highest quality rating in the industry.

Other Positions Held: Personnel Manager, Purchasing Manager, Manufacturing Engineer

EDUCATION

Illinois Polytech — Electrical Engineering
Evanston, Illinois

Urbana Community College — AA Degree - General Education
Urbana, Illinois

JOHN C. DOE

1072 Somewhere Avenue
Anywhere, Michigan 48084
(313) 748-7849

PROFESSIONAL HISTORY

Tonawanda Corporation, Medina, Michigan
July 1958 to present

PRODUCT MANAGER

Responsible for the operation, administration and overall profitability of an $80,000,000 per year product line. Developed and implemented annual product plans. Directed international and national sales managers on methods to maximize profitability.

As chairman of the management committee, supervised ten high-level managers charged with the responsibility of reducing costs, improving sales, and maximizing profits.

-Improved profit of product line $6,000,000 in 1968.

-Implemented successful transition from four production plants to one.

-Initiated programs which resulted in a 20% reduction in production costs.

DISTRICT MANAGER

Responsible for sales and customer relations for all products to original equipment manufacturers in a six-state district, headquartered in Atlanta, Georgia.

-Increased annual sales from $3,500,000 to $4,500,000 within 18 months.

TERRITORY MANAGER

Responsible for expanding existing dealers and establishing new accounts in the Fort Worth, Texas area.

-Increased annual sales from $500,000 to $1,600,000.

-Increased dealer locations from 4 to 21.

Feistan Consulting, Inc., 245 Someplace St., Anywhere, MI

DIRECTOR
(June 1980 to present)

Member of the Board of Directors of one of the most respected tire-design consulting firms in the industry.

EDUCATION

University of Detroit
Bachelor of Business Administration, Marketing

PERSONAL:

Married * 4 Children * Excellent Health
Willing to Travel and/or Relocate

GILDEA F. EWING

2837 New Hudson Road
Woodlake, Ohio 38745

PROFESSIONAL HISTORY:

Autotech, inc., 1827 W. 15th Ave, Brighton, Ohio
1986-Present

SECRETARY TO THE CONTROLLER
Provided administrative and secretarial support to the Controller of a major supplier of automotive engineering and design services. Frequently interfaced with department managers and their secretaries at the administrative headquarters to coordinate mutual projects. Position requires a high degree of confidentiality.

- Preparation of financial statement binders for distribution to executive staff.
- Assisted controller in development of spreadsheets.
- Typing various reports utilizing Text Management and IBM Smart II.
- Generated productivity graphs and reports using IBM System 38.
- Assisted Payroll Department in calculating and posting time cards utilizing the Software 2000 System. Also assisted Accounts Payable Department with computer-generated check runs.
- Prepared bank deposits, disbursing and balancing petty cash.
- Arranged luncheons and meetings for up to 150 people.
- Handled routine correspondence and memos, screened and answered phone calls.

COST ACCOUNTING CLERK
Processed time sheets, entered data and generated weekly reports using IBM System 46. Reconciled shipping documents with Order Desk documents. Checked current and standard indented Bill of Materials for billing purposes.

First of America Bank, 4738 East 12 Mile, Southfield, MI
1981-1986

ACCOUNTING CLERK/TYPIST
Prepared checking account statements. Informed managers of overdrawn accounts. Processed returned items, microfilmed daily work, and served as back up to bookkeeper. Handled customer telephone inquiries.

EDUCATION:

PC Based Applicants Course	Smart II
Lake County Community College	Medical Terminology Basic Business, Computers
Arlington High School Indianapolis, IN	Diploma Business Major

CATHERINE O'HENRY

3453 East 58th Street
Madison, Wisconsin 54857
(608) 389-3498

EMPLOYMENT HISTORY:
General Manufacturing, Inc., 7837 W. Illinois Ave, Madison, WI
January 1978 - Present (Available due to plant closing.)

ADMINISTRATIVE SECRETARY
Responsibilities involved all current office procedures and the implementation of new procedures designed to enhance efficiency of plant operations. Performed secretarial duties and special projects which required a high degree of confidentiality. Frequently interfaced with department managers and their secretaries at the administrative headquarters to coordinate mutual projects.

Other duties included:

- Maintaining personnel files, time cards and payroll for all hourly employees.

- Processing all medical reports for on-the-job injuries and directing insurance forms to proper channels.

- Typing, processing and expediting of all purchase orders.

- Disbursement of petty cash and approval of expenditures.

- Handled monthly cost accounting reports.

- Monitored work in-process and close-out of completed jobs.

- Inventory control and customer billing.

- Developed filing system and work-flow system for the entire plant.

* Enjoys problem-solving and is constantly willing to take on additional responsibilities and challenges.

RECEPTIONIST
Responsible for a variety of receptionist and clerical duties for a computer service company specializing in tax processing. Screened and directed phone calls, greeted visiting customers, sorted and directed mail. Typed reports, correspondence and labels on word processor. Processed billing and assembled mailings. Entered client information into list processing program. Sorted and printed labels for mailings.

- Established an excellent rapport with customers and co-workers.

- Commended for efficiency and attention to detail.

EDUCATION: Valley View Community College, Madison, WI
Currently enrolled in computer programming.

HOBBIES: Horses, water sports, reading and sewing.

LINDA P. ENGLES

98 Seventh Avenue
Oaklawn, Wisconsin 54834
(608) 746-9304

Skilled, Certified Secretary, capable of working with little supervision. Ability to prioritize and produce results in a busy work environment.

PROFESSIONAL EXPERIENCE:

GHC, Co., Oaklawn, Wisconsin — 1987-Present
(Available due to corporate restructuring.)

SECRETARY/RECEPTIONIST
Responsible for a variety of receptionist and clerical duties for a computer service company specializing in tax processing. Screened and directed phone calls, greeted visiting customers, sorted and directed mail. Typed reports, correspondence and labels on word processor. Processed billing and assembled mailings. Entered client information into list processing program. Sorted and printed labels for mailings.

- Demonstrated proficiency with word processing system and Xerox Memory typewriter.
- Established an excellent rapport with customers and co-workers.
- Commended for efficiency and attention to detail.

J.P Luds, Inc., Detroit, MI — 1974-1987

CASHIER SUPERVISOR
Responsible for training cashiers, writing performance reviews, monitoring cashier schedules, delegating work, operating store computer and assisting customers.

EDUCATION:

Ramsey State Junior College
Ramsey, Michigan — Secretarial Certificate

HOBBIES:

Bowling * Golf * Photography

MEMBERSHIPS:

P.T.A. - Treasurer, 1987-1988

References Available Upon Request

ELLEN S. FRASER

3241 Louise Avenue
Sherman, Michigan 48123
(313) 649-1576

Professional History:

CONFIDENTIAL SECRETARY TO PLANT MANAGER

American Bolt & Screw Company
415 Remington Boulevard, Detroit, MI 48707
1978 - 1990

Performed secretarial duties and special projects which required a high degree of confidentiality. Typed employee status notices and salary reviews, established travel itineraries for employees sent on field service calls, collated weekend overtime for all departments, calculated and charted plant efficiency, productivity and utilization. Frequent contacts with department managers and their secretaries to establish meetings and agendas.

- Considered by management to be loyal, dependable and responsible.
- Cited by management for cost improvement suggestions which saved company $33,000.
- Received certificate for completion of a computer-assisted program to enhance secretarial skills.

PERSONNEL MANAGER

Mapsco Manufacturing Company, 3404 South Haven Street, Chicago, IL 60607
1973 - 1978

Maintained personnel files on up to 200 employees. Handled payroll, including timecards, stamping and distribution of pay-checks, deductions for wage assignments, etc. Responsible for all business correspondence.

Education:

Attended Bay College, Delta, Michigan
Completed coursework in Business Administration

Graduated from Salina High School, Salina, Michigan
Diploma in Business

Personal Data:

Excellent health
Willing to travel or relocate

CLEMENTINE C. PETERSON (313) 769-0340
9394 Stephenson
Troy, Michigan 48408

SUMMARY:

Highly motivated achiever with outstanding credentials and in-depth experience involving secretarial, administrative, analytical and coordination skills. Goal and team oriented, recognized as working well with both peers and management. Dependable, possessing strong work ethic, and adaptability to change.

Computer Skills:
- Proficient in WordPerfect 5.1, Display Write III & IV, Lotus 1-2-3, DOS and related programs.
- Developed and maintained monthly income statements, labor reports, and computerized accounting procedures for three major firms.

Clerical Skills:
- Typing - 90 wpm accurately; shorthand 120 wpm accurately; Dictaphone transcription, operate variety of office equipment; excellent spelling, punctuation, mathematical skills.

MAJOR ACCOMPLISHMENTS:

- Assisted executive producers, directors and writers in execution of Smith Corporation's New Car Announcement Shows for three consecutive years. Interfaced closely with senior management of Smith Corporation. Specialized in on-location office set-up and management of all phases of the project.
- Effectively met computer input deadlines of business meeting speeches for Smith senior management.
- Developed and implemented an informational manual for distribution to all Jones Motor Division dealerships for use in their day-to-day computerized accounting programs.
- Trouble shooter for the Jones Dealer Network. Assisted dealers when electronically submitting monthly financial reports through TIMS (Total Information Management System) and various IMS programs developed by EDS specifically for Aberdeen Co.
- Supervised and established initial start-up of electrostatic painting operation, including screening, interviewing and orientation of new employees. Assisted with development of policies and procedures handbook for new employees.

EMPLOYMENT HISTORY:

1987 - Present	Self Employed Accountant
1989 - 1991	Kelly Productions Senior Secretary/Office Manager
1984 - 1988	Manpower (contracted to Jones Motor Division) Analytical Coordinator/Senior Word Processing Specialist

Prior Experience: Secretary, Stenographer, Receptionist

EDUCATION:

Oakland County Community College, Business
Ferndale Adult and Community Education, Accounting
Smith Motors Corporation - Professional Training Courses:
- Time and Self Management
- Stress Management
- Professional and Personal Excellence
- Equipment and Software Training

ROBERTA P. RUSSELL

33095 Highland Drive, Apt #4
Sterling Heights, Michigan 48305
(313) 467-5553

Highly motivated employee with outstanding credentials and over 20 years' experience involving secretarial, administrative, and coordination skills. Goal and team oriented, recognized as working well with both peers and management.

Computer Skills:
Proficient in WordPerfect 5.1, DisplayWrite III & IV, Lotus 1-2-3, DOS and related programs. Developed and maintained monthly income statements, labor reports, and computerized accounting procedures for three major firms.

Clerical Skills:
Typing - 90 wpm accurately; shorthand - 60 wpm accurately (capable of up to 120 wpm); Dictaphone transcription; operate variety of office equipment; excellent spelling, punctuation, mathematical skills.

MAJOR ACCOMPLISHMENTS:

- Assisted executive producers, directors and writers in execution of Chrysler Corporation's New Car Announcement Shows for three consecutive years. Interfaced closely with senior management of Chrysler Corporation. Specialized in on- location office set-up and management of all phases of the project.

- Effectively met computer input deadlines of business meeting speeches for Chrysler senior management.

- Developed and implemented an informational manual for distribution to all Chevrolet Motor Division dealerships for use in their day-to-day computerized accounting programs.

- Supervised and established initial start-up of electrostatic painting operation, including resume screening, interviewing and orientation of new employees. Assisted with development of policies and procedures handbook for new employees.

EMPLOYMENT HISTORY:

1986 - 1991 — Yednock Productions
Senior Secretary/Office Manager

1982 - 1986 — Norell Services (contracted to Chevrolet Motor Division)
Analytical Coordinator/Senior Word Processing Specialist

1975 - 1981 — LeClare Color Coat, Inc.
Personnel Director/Executive Secretary

EDUCATION:

Macomb County Community College, Business

Troy Adult and Community Education, Accounting

General Motors Corporation - Professional Training Courses:
Time and Self Management — Professional and Personal Excellence
Stress Management — Equipment and Software Training

BREE A. LUTZ
3874 South Durand Road
Union Lake, Michigan 48386
(313) 372-9309

EMPLOYMENT HISTORY

Automotive Dynamics, Inc. 1976-1991
Westland, Michigan

Administrative Assistant
Provided administrative assistance to the President and Senior Executive staff of a major specialty vehicle manufacturer. Directed activities including: preparation and transcription of correspondence and corporate reports, utilization of IBM System 38, Text Management and Lotus 1-2-3, scheduling of appointments and travel itineraries, screening incoming correspondence and calls, and other current office procedures. Additional administrative responsibilities and accomplishments include:

- Evaluated, recommended, managed and maintained a cost-effective corporate telecommunication system.
- Coordinated all corporate travel (including satellite plants); implemented agency revenue sharing plan and hotel/auto rental direct billings that realized significant corporate savings of 10% annually.
- Initiated, negotiated and oversaw a unique office supply inventory concept that resulted in up-front corporate savings.
- Supervised engineering reproduction, mail and switchboard operations.
- Coordinated and planned special corporate meetings and event, including annual Auto Expo.
- Recognized by senior management as quality conscious, highly motivated, with strong problem solving and decision making skills.

ADDITIONAL SKILLS AND EXPERIENCE

Secretary for a Marine Surveying office, a subsidiary of Lloyds of London. Duties included: typing, dictation, transcription, filing, teletyping, researching and collating survey report information. Frequent contact with the U.S. Coast Guard and client agencies.

EDUCATION

Wayne County Community College — Pursuing a degree in Business Administration
Detroit, Michigan

Oakland Community College — Data Processing (1 year)
Farmington Hills, MI

VOLUNTEER WORK

Board Member - Walled Lake School District
Girl Scouts of America - Leader - 5 years

JEANNE A. WESTMAN

3895 Wing Lake Road
Holly, Michigan 48435
(313) 389-3905

Highly professional administrator with strong organizational skills and extensive background in both business and educational environments.

Professional History

EXECUTIVE/ADMINISTRATIVE ASSISTANT 1985-1991

BCA Merchandise, Inc.
Rochester, Michigan

Assistant to Chief Financial Officer, Vice President of Administration and Operations, and National Controller. Handled all correspondence, telephones, weekly aging reports on national accounts; maintained certificates of product liability for all divisions; generated and distributed monthly, quarterly and annual financial statements; maintained files; scheduled airline reservations for executives at location; and reconciled American Express statements with expense reports.

Proficient on WordStar and WordMarc (Primeword) word processing systems, as well as with Lotus 1-2-3 and 20/20 (Prime) spreadsheet programs.

Type 70+ words per minute and proficient in all office equipment.

SECRETARY TO SCHOOL PRINCIPAL 1981-1985
Columbus Board of Education, Columbus, Ohio

Typed correspondence; maintained student records; generated weekly parent newsletter; communicated with parents; ordered supplies; and kept inventory records.

TRAINER/CONSULTANT 1977-1981
Indiana Bell Communications, Indianapolis, Indiana

Conducted corporate training of business telephone systems in Toledo area and training session for personnel on new telephone equipment after installation.

CLASSROOM TEACHER 1972-1975
Birmingham Public Schools, Birmingham, Alabama

Taught two years (all subjects) at 5th grade level and one year at 6th and 8th grade levels in language arts and history.

Previous employment: Receptionist, Sales Clerk, Teacher's Aide

Education

Word processing and spreadsheet training through Cogent Information Systems and Prime Computer.

Bachelor of Arts
Ohio State University

Language Arts major with minors in Secondary Education and Psychology.

Volunteer Work

Designed and conducted seminars on telephone skills for various non-profit organizations.

ALEXANDRIA C. POLCHIN

38411 East Edinborough Court
Madison Heights, Michigan 48240
(313) 795-2673

Highly skilled professional with strong background in the areas of Personnel Administration, Benefits and Compensation, and Office Management.

EDUCATION: Eastern Michigan University
Master Science, Administration, 1989

Eastern Michigan University
B.S. Management & Supervision, Magna Cum Laude, 1982

Key Courses:
Financial Analysis
Quantitative Applications for Decision Making
Strategic Planning Management
Public Relations Management
Marketing Management
Consumer Behavior
Advertising Management
International Marketing
Personnel Administration

PC SKILLS: Lotus, Harvard Graphics, Statgraphics, dBASE, Microsoft Word, WordPerfect, and DisplayWrite.

BUSINESS EXPERIENCE:

1976 - Present
Liberty Division, Dillman Industries Inc.

Administrative responsibilities in Group Headquarters of major manufacturer of automotive components. Participate in administrative components of acquisitions, mergers, long-range planning, market development, budgeting, and financial analysis.

General Administration. Monitor and track capital and expense projects for divisions. Prepare and monitor Dillman group office budget. Approve and oversee expenses for Dillman office group. Organize, plan and participate in executive meetings; negotiate with meeting sites, purchase meeting materials. Provide administrative support to Group Vice President.

* Formulated, developed, and computerized program (Lotus 1-2-3) to consolidate and summarize monthly divisional financial information.
* Synchronized the design, construction, and decoration of group office relocation.

Personnel Administration. Maintained human resources information system and salary records for executive staff. Administered benefits program for the Group; authorized health insurance claims; prepared life, health, and disability premium statements; counseled employees re: benefits issues; liaison with health care providers.

* Developed and implemented computerized employee benefits information system. The new system resulted in more accurate premium payments, annual individualized benefits statements, and more effective benefits analysis.
* Developed COBRA program; employee notification, billing methods, and adherence to current legislation.

Prior Assignments: Started with Liberty as Executive Secretary. Provided secretarial support to Vice President, Marketing and Vice President, Advanced Engineering.

LAURA N. PALMER

100 Walker Avenue
Priceville, Ohio 94783
(320) 647-9045

EMPLOYMENT EXPERIENCE:

Textell, Inc., 1332 North Smith, Priceville, Ohio 94783
1978 - 1991
(Due to product line phase-out.)

PERSONNEL CLERK

Employed by a major manufacturer of computer peripherals. Responsible for interview screening, hiring, maintaining employment records and handling terminations, sick-leave forms and vacations.

Published and designed quarterly office newsletter.

SWITCHBOARD OPERATOR/RECEPTIONIST

Responsible for all switchboard and front desk duties. Experienced with both Northern Telecom and Ohio Bell systems.

BLUEPRINT CLERK

Operated a Xerox Blueprint machine and Kodak copier in the Engineering Department. Also responsible for changes, additions, binding and ordering instruction manuals.

- Excellent work and attendance records.
- Received written recommendations from management.
- Always received above average performance evaluations.

EDUCATION:

Priceville College Priceville, Ohio	Office Administration Associates Degree

HOBBIES AND INTERESTS:

Volunteer work, gardening and golf

PERSONAL DATA:

Married, excellent health

References Furnished Upon Request

FIELD A. SEYFORTH

8342 Constitution Avenue
Madison Heights, Michigan 48206
(313) 643-9483

Professional, dependable , hard-working, detail-oriented individual. Experienced with MultiMate Advantage II, Lotus1-2-3/Always, and Lotus Freelance. Type 65 wpm, and possess excellent communications skills and a pleasant phone manner.

WORK EXPERIENCE:

XYZ-Vehicle Safety Systems, Inc., Arlington, MI — 1987-1991
(available due to corporate down-sizing)

Purchasing Clerk
Provided clerical support including: entering purchase orders, filing, typing correspondence, memos and other documentation. Worked with McCormack & Dodge Software, Lotus/Always, Freelance, and MultiMate Advantage II. Answered telephone, developed presentation materials, maintained daily calendar of events and scheduled meetings.
Analyzed and developed new filing system procedure and coordinated new related security system.

Cost Accounting Clerk
Involved in data entry on IBM computer system. Typed letters, memos and various other documents. Worked on bank reconciliations and functioned as switchboard operator.

File Clerk - Accounts Payable Department
Handled filing, processed invoices, and performed general office work including operation of the switchboard.

Clerk - Human Relations Department
Performed general office procedures. Set up new filing system, and maintained employee vacation and sick leave records.

EDUCATION:

Oakland Community College, Center Campus — 1987 - Present
Currently pursuing an Associates in General Business, to be completed in December 1991.

Courses of special interest:

Accounting I, II	Business Management
Economics I, II	Marketing
Program Development	Business Law I
Computers	Cobol Programming

HOWARD J. WOLFGANG

375 Lussier Drive
Santa Ana, California 98776
(805) 952-9351

Experienced professional security manager with strong law enforcement background.

Work History:

The United States Manufacturing Company, Long Beach, CA
(1980 to Present)

CHIEF SECURITY OFFICER
Direct a staff of 25 security officers and manage daily security operations for a 150-acre manufacturing site, including plants, warehouses, yard, entrances, rail spurs and parking lots. Conduct investigations of internal and external thefts and other security violations. Maintain fire protection systems, plant sewer effluent operations and medical department on weekends. Acting Security Director in his absence.

- Developed and implemented a new Detex security key route, improving security officer efficiency and overall site security.
- Devised a method to achieve early detection of break-ins resulting in an 60% reduction of incidents.
- Introduced a plan to reduce vehicle traffic in and out of the site, resulting in greater warehouse security and reduction in thefts.
- Recognized by management for high standard of integrity, sense of responsibility and the ability to handle and resolve conflicts.

State of California, Department of Law, Office of the Attorney General
Los Angeles, California (1975 - 1980)

SENIOR CONFIDENTIAL INVESTIGATOR
Developed, implemented and administered methodologies to successfully deal with white collar crime, including investigation, arrest and conviction. Directed a staff of 12 investigators, using sophisticated electronic equipment to investigate securities fraud, stock and bond manipulations, environmental abuses, and civil rights violations.

- Achieved 98% conviction rate on arrests
- Commended by Attorney General's office for successfully handling several significant cases

Other Experience

Los Angeles City Detective

Owner of Wholesale Bakery, Pasadena, California

Education

California State University, Los Angeles, CA
Master of Social Science - Criminal Justice

City University of Los Angeles, Los Angeles, CA
Bachelor of Arts - Sociology

Interests & Activities

Golf, running, tennis, active in civic organizations.

BOBBY M. LEWIS

1423 West Outer Drive
Hillsdale, Michigan 48513
(517) 893-9403

PROFESSIONAL EXPERIENCE:

ASSOCIATE TECHNOLOGIST
American Foods, Inc., Hillsdale, Michigan
1985 - Present, (due to downsizing).

Responsible for executing technical experiments using established laboratory procedures. Interpret results and translate into possible solutions to problems. Ability to follow directions on objectives and report results within established timeframes and other parameters.

Maintain accurate, complete research records. Seek, evaluate and utilize technical information from colleagues and relevant literature as appropriate to achieve experimental objectives. Propose and formulate detailed experimental procedures within specific project objectives.

Able to recognize, identify and communicate significant or unexpected research findings.

Demonstrated technical competence by contributing to the development and implementation of key product improvements, cost reduction and new product and line extensions.

SKILLS:

* Considered by management to have excellent written and verbal communication skills.

* Recognized as self-motivated, creative, and cooperative.

* Commended for exceptional problem-solving and troubleshooting ability.

EDUCATION:

Michigan State University
East Lansing, Michigan
B.S., Major: Chemistry, Minor: Math

MILITARY HISTORY:

United States Marine Corps
Staff Sergeant - Honorable Discharge

HOBBIES AND INTERESTS:

Sailing * Golf * Woodworking

PERSONAL DATA:

Excellent health * Married * Three children

References available upon request

WILLIAM R. SMITH

5414 Dante
Detroit, Michigan 48210
(313) 457-0330

CHEMIST with in-depth experience in compounding and formulating SEALERS and ADHESIVES. Developed products from laboratory stage through pilot plant, to actual production for automotive manufacturers including: General Motors, Ford, Mazda, and Honda.

EXPERIENCE

1977-1991, TROY SPECIALTY PRODUCTS
(a subsidiary of White Chemical Corporation)

Chemist

Responsible for research and development of sealants and adhesives to meet automobile manufacturers' specifications. Developed formulas, processes, techniques, and methods... all steps from small laboratory through large quantities. Conducted material tests including: salt sprays, modulus, tensile strength, weldability, paintability, etc. Prepared detailed reports on each project, utilizing IBM PC.

- Developed and researched various cold pumpable expandable sealers, from laboratory status to production status ($5-10M).
- Developed extruded hot melt sealers. Currently in G.M. and Ford facilities ($2-3M).
- Each year conducted a Safety Seminar on correct, safe operation of all the mixers and mills in the laboratory.
- O.S.H.A. and M.S.D.S. in-house training. Supervised and implemented safety in the laboratory.
- Worked closely with Japanese firm to replace Japanese raw materials with domestic sources.
- Supervisor of Pilot Plant at Warren Facility. Supervised several employees for pilot production orders.

Prior Experience: Lab Technician

EDUCATION

Lawrence Technical University
Concentration in Polymer Chemistry and Mechanical Engineering

Company sponsored courses and seminars including:

- Society of Automotive Engineers Seminar
- Rubber Compounding and Blending
- Analyzing Organic and Inorganic Compounds
- Monthly open discussion with peers and other group leaders on problem solving for specific products

LYN B. JOHNSON

4382 Elk Ridge Road
Cleveland, Tennessee 37311
(615) 783-4095

EMPLOYMENT HISTORY:

LABORATORY TECHNICIAN
American Food Products, Inc. 5483 Maple Drive, Cleveland, Tennessee 37311, (available due to plant closing).
1977 - Present

Responsible for performing essential analytical testing on various commodities and ingredients used in the processing and production of grain products. Duties include: taking, collecting, and preparing samples for testing. Maintain accurate records. Write required reports and prepare and use sprays and fumigants.

* Supervised and trained entry-level technicians.

* Commended by management for accuracy and attention to detail.

* Excellent attendance and work record.

GENERAL LABOR - TESTING
United Foods, Inc. 4783 Maple Drive, Cleveland, Tennessee 37311
1974 - 1977

Performed quality assurance tests as assigned, completing necessary forms and maintaining required records to assure accurate monitoring of quality procedures in blending and mixing operations. Also completed mathematical and written reports.

* Increased accuracy in mathematical reports by 13%.

* Promoted to Laboratory Technician after only three years.

Additional positions have included: file clerk, shipping and receiving clerk, receptionist

EDUCATION:

Pine Valley High School
Pine Valley, TN
Graduated

HOBBIES AND INTERESTS:

Fishing * Woodworking * Camping * Golf

References Available Upon Request

ROBERT J. RIVERSIDE

2152 Cheshire Road
Mapleton, Michigan 47867
(313) 987-9078

EMPLOYMENT HISTORY:

Amax Industrial, Inc., 4382 RR#3, Mapleton, MI 47865,
June 1973 - Present, (available due to plant closing).

ELECTRONIC TECHNICIAN
Responsible for a variety of duties necessary to electrical/electronic fabrication process for a major manufacturer of high quality machine tools.

Supervise and train new personnel, complete all necessary documentation and provide ongoing support and guidance through special project completion.

Qualified in the following areas: machine and equipment installation, service and repair.

Skilled at trouble-shooting and devising most expedient repair procedure. Design, lay-out and fabricate special parts and equipment.

* Excellent written and oral communication skills.

* Commended by management for leadership.

* Outstanding attendance and safety records.

* Accurate, reliable and hardworking.

MAINTENANCE HELPER
Assisted in the installation and maintenance of machinery and equipment. Able to work independently performing routine trade functions such as: rigging block and tackle, disassembly of machinery and equipment, measuring, cutting, threading, and forming and fitting.

* Perfect attendance 4 years in a row.

* Trained and supervised other maintenance helpers.

MILITARY HISTORY:

U.S. Navy — Stationed in Providence, RI
Trained in machine repair

EDUCATION:

Associates Degree - Electronics
Oakridge Community College
Oakridge, MI

HOBBIES AND INTERESTS:

Restore, buy and sell automobiles

LYNN P. HUNTINGTON

3843 Monroe Avenue
Battle Creek, Michigan 49395
(616) 348-5816

Employment Experience
Clark Company, a division of Pinicle Data Systems, Battle Creek, MI
1983 to Present

DETAIL DRAFTER
Solely responsible for all drawings of electrical and mechanical parts for a manufacturer of electronic products. Worked from sketches, verbal instructions and sample parts to create or revise appropriate drawings. Interfaced with mechanical and electrical engineers to check finished drawings. Processed and typed mini-prints with the Wordstar word processing system, compiling mechanical drawings, schematics and written instructions into report form for perusal by engineers. Organized and maintained the company's filing system for the drafting department. Experienced with both Cadkey CAD drafting system as well as pencil sketches.

Established a library of standard notes for the Cadkey CAD drafting system which organized the guidelines and specifications for various parts manufactured by Clark.

Responsible for separating all Clark blueprints from the parent company's prints which resulted in quicker revisions of drawings.

Chosen for position as sole Clark drafter based on knowledge of Clark products, skills, and positive attitude toward work.

Constantly willing to acquire new skills through training to improve job performance.

Started in company as manufacturing process clerk, processing paperwork for production lines. Served as liaison between manufacturing engineers and production supervisors to ensure engineering changes were made.

Entech Placement Service, Davisburg, MI - 1983

CAD DETAIL DRAFTER
Detailed drawings from sketches and verbal orders of mechanical engineers on CADDS 4X REV.3 system. Worked with designers on parts for prototypes of electrical appliances while under contract for Hotpoint. Experienced in creating original drawings from personal ideas.

Previous Work Experience: Bookkeeper, Freelance illustrator, Engineering Clerk

Education
Lake Michigan College, Benton Harbor, Michigan
Associates Degree in General Studies majoring in CAD drafting.

Purdue University, West Lafayette, Indiana
Two years study in the field of art and design.

Special Classes

Business Accounting
Advance Wordstar
Materials Management
Business Correspondence
AC/DC Electronics
Graphic Art
Lotus 1-2-3

Volunteer Work   Girl Scouts of America - Leader

L O N C. D E V I N

5385 Moss Court Lake Orion, Ohio 48362 (303) 933-2124

QUALIFICATIONS

* Hands-on manager with excellent technical and interpersonal skills. Ability to communicate and interact with a wide variety of individuals and groups.

* Experienced in laboratory development, data analysis, production implementation of innovative materials and procedures, and constructing improvement plans.

* Directed quality assurance activities involving over $750 million annually in raw materials.

* Well-developed and extensive knowledge of raw materials, rubber compounding, rubber coated fabrics/wire, and suppliers of elastomers, wire, fabrics and chemicals.

* Experienced Quality Systems Auditor, knowledgeable about ISO 9000, Targets for Excellence, and Q1 requirements, expert in supplier and customer site auditing.

ACCOMPLISHMENTS

* Consistently selected for progressively more responsible positions, recognized for technical expertise and quality assurance excellence.

* Invented and developed new compounds, resulting in improved major products and reduced costs.

* Developed textile adhesive dip for polyester fabric that became an essential component for a major product.

* Developed innovative procedures and systems to monitor raw material quality, vendor compliance and warn of substandard.

* Selected to lead combined companies' quality program. Established the merged and highly effective quality assurance department.

PROFESSIONAL EXPERIENCE

B.F. Goodman Tire Company, Troy, Ohio 1981 - Present
SENIOR RESEARCH ASSOCIATE, Raw Materials, Specifications, Processing and Reinforcements
Manage and direct raw material activities including: raw material development, specifications, approval and evaluation; profit improvement programs. Acted as supplier technical liaison.

MANAGER, Raw Materials Quality Assurance
Promoted to this role to manage quality activities of purchased raw materials for the newly-merged Unitire and Goodman companies; including material compliance, supplier performance monitoring, and customer quality requirements.

RESEARCH SCIENTIST, Materials Compounding
Performed technical research resulting in the invention and development of new tire compounds.

Grimstone Tire and Rubber Company, Akron, Ohio 1974 - 1981
Hired as Adhesives Compounder and promoted to Senior Compounder after three years.

PROFESSIONAL CERTIFICATIONS/AFFILIATIONS:
American Society for Quality Control:
Certified Quality Engineer
Certified Quality Auditor

EDUCATION University of Michigan, Ann Arbor, Michigan
Bachelor of Science, Chemistry

RICHARD P. BANNEY

584 Sidney Circle
Flint, Michigan 48502
(313) 653-4894

Master Plumber with diversified supervisory and field experience.

WORK HISTORY

FOREMAN
Lester Mechanical, Plymouth, Michigan 1981-1992

Direct and schedule the activities of up to 15 apprentice and journeymen plumbers involved in large commercial projects over a 100 mile radius. Responsible for continuous work-flow and overall quality. Maintain progress reports and inventory requirements. Coordinate with other contractors' schedules and oversee sub-contractors. Additional duties include enforcing safety rules and demonstrating time-saving and labor-saving techniques.

* Foreman on Hilton Hotel project; involved in layout, installations and supervision of entire project. ($1M).
* Foreman on Novi School's three major renovations. ($500,000+)
* Co-Foreman on Oakland University installation of new laboratories. Supervised installation of acid waste, stainless steel piping, deionized water system. ($450,000)
* Co-Foreman on Birmingham School's renovation of 25 classroom heating & cooling unit ventilator. Installed chilled water system (45-ton), and new domestic hot water boiler. ($600,000)

JOURNEYMAN PLUMBER
Osgood Plumbing, Pontiac, Michigan 1976-1981

Performed all types of commercial, industrial and residential plumbing work. Involved in design, installation and testing of new systems. Qualified blueprint reader. Trained new employees working toward journeymen status.

PREVIOUS EMPLOYMENT

Wholesale Plumbing Sales, General Construction, Landscaping and Maintenance

EDUCTION

Completed Plumbing Apprenticeship - Licensed

Notre Dame High School
Detroit, MI
Graduated

HOBBIES AND INTERESTS

Boy Scout Leader, Camping, Fishing

DONNA R. CAMPBELL

2121 Trail Road
Troy, Pennsylvania 46484
(317) 856-0349

EMPLOYMENT HISTORY:

Troy, Inc. 3874 Maple Rd., Troy, PA 46484
1974 - Present, available due to permanent plant closing.

MAINTENANCE/MATERIAL HANDLER
Responsible for maintaining production banks and securing and moving supplies, materials, waste, or equipment to and from designated areas for a major manufacturer of small engine parts.

Skilled at performing a variety maintenance duties required to keep work areas safe and clean.

Trained in the proper and safe handling of hazardous wastes.

Maintained accurate records on the lubrication of all mechanical equipment. Ordered supplies when required.

Operation of the following:

Fork-lift truck   Hand and power tools   Air Compressor

* Excellent work and attendance records.

* Perfect safety record for 10 years in a row.

* Commended by supervisors for self-motivation and problem-solving.

Other work experience included:

Short order cook	Delivery truck driver	Farm work
Groundskeeper	Janitor	Auto body work

EDUCATION:

Currently enrolled in Adult Education Program, "Introduction to Computers".

In-house training completed: Safety and the Handling of Hazardous Wastes.

Maplevale High School
Maplevale, PA

HOBBIES:

Gardening and volunteer work

References Available Upon Request

BRIAN L. MARKS

367 Central Street
Dow, West Virginia 32645
(304) 521-0230

EXPERIENCE

1979 - 1992

Great American Mining Company
Dow, West Virginia

MINER
Performed combination of tasks utilizing high-voltage electric-powered equipment. Cut channels under working face to facilitate blasting, operated mounted power drill to bore blasting holes in working face, charged and set off explosives to blast down coal. Installed timbering and roof bolts.

* Perfect safety record.
* Excellent attendance record.
* Selected by management to be a participant in Great America's Total Quality Productivity Planning Team.

Other Experience:

Seasonal as self-employed carpenter: installed windows and cabinets, built garages, installed dry wall. Experienced in rough and finished carpentry.

Part-time work as self-employed small engine repairman: repaired and rebuilt mowers, outboard motors and chain saws.

EDUCATION:

Dow Vocational-Technical School
Graduated with honors.

COMMUNITY ACTIVITIES:

Volunteer literacy tutor
Sunday school teacher
Secretary, Dow Community Beautification Commission

YARDLEY F. DONEREZ

RR 6 Box 3894
Meadow Lake Dr.
Fenton, Pennsylvania 17332
(716) 346-3331

EMPLOYMENT HISTORY: MACHINIST

Hi-Drive, Inc.
1975-1987
Terre Haute, IN
1987-1991
York, Pennsylvania

(Available due to permanent layoff)

Responsible for selecting the proper tools to set up and operate a Greenlee screw machine utilized to process and machine various heavy equipment parts for a major international manufacturing firm producing heavy duty engines, construction and farm implements.

- Employee Satisfaction Process Group Leader
- Employee Group Leader, Training
- Quality Circle Group Member
- Qualified blueprint reader
- Designed special tools when needed to do salvage and rework
- Suggested and implemented methods to eliminate one complete drilling operation, resulting in a savings of $65,000 annually
- Changed a process from company layouts which created 40% scrap to a more efficient method, cutting scrap to 5%

OTHER:

- Liaison to the U.S. Army for Maple Tree Gun Club. File papers interpreting army regulations to set up marksmanship programs for the club. This included budgets, obtaining materials, writing rules, overseeing and training volunteers
- On Board of Directors of 2 gun clubs
- Computer literate
- Management Trainee; supervised 12-person department periodically

EDUCATION

St. Francis University
Terre Haute, In
Elementary Education

Lane Community College
Indianapolis, IN
Manufacturing Technology Course

York Vo-Tech
York, PA
Hi-Drive Skills Upgrade

PERSONAL DATA: Married * 3 children * Health excellent

GARY P. RICHMOND

3214 Stanley Street
Indianapolis, Indiana 49203
(317) 793-0398

EMPLOYMENT HISTORY:
Continental Camshaft, Inc. 234 Maple Rd., Indianapolis, IN 49203
1979 - Present

MACHINE REPAIRMAN
Responsible for building, repairing, replacing and assembling production and machine shop equipment for a major supplier of automotive camshafts.

Plan work sequences, diagnose machine malfunctions and perform repairs necessary to maintain production schedules. Skilled at machining simple replacement parts. Familiar with the accurate use of scales, calipers, micrometer, scribers, various gauges and other precision instruments. Maintain records on all work performed.

* Outstanding trouble-shooter, work well under pressure of deadlines.
* Accurate, well-organized worker, self-motivated.
* Perfect attendance for last 5 years.
* Excellent safety record.

MILITARY HISTORY:

U.S. Navy   Honorable Discharge   Machine Repairman

EDUCATION:

Continental In-House Safety Training - 1979-1991

Arlington High School Indianapolis, IN	Graduated Industrial Arts

HOBBIES:

Bowling, golf and fishing

PERSONAL DATA:

* Excellent health   * Three children   * Willing to travel

References Available Upon Request

MICHAEL J. NODELL

1432 Main Street
Nashville, Tennessee 37202
(615) 897-0988

EMPLOYMENT HISTORY:

American Products, Inc., 1728 Hill St., Moline, IL 79837
1984 - Present, (due to plant closing).

BOILER OPERATOR
Responsible for controlling the operations of four boilers (capacity - 200,000 lbs/hr.) and related auxiliary equipment necessary for providing steam for process and heating for the plant.
Possess thorough knowledge of the operation, control and construction of boilers, pumps, air compressors, temperature and pressure gauges, safety and relief valves and other miscellaneous boiler house equipment.
Certified by the state and possess a third class engineer's license

* Excellent attendance and safety record.

* Commended by supervisors for outstanding job performance.

ASSISTANT BOILER OPERATOR
Trained in all aspects of boiler operation. Promoted to Boiler Operator after one year.

U.S. Travelall, Inc., 1534 Maple Ave, Moline, IL 79837
1979 - 1984, (due to plant closing).

ASSEMBLER
Member of a production team engaged in the assembly and manufacture of high-quality heavy equipment components. Responsible for the assembly and installation of standard and special piece-parts and sub-assemblies for subsequent assembly to the prime product. Familiar with the operation of a variety of hand tools and power tools including the following: hydraulic press, drill press, and rivet squeeze. Performed inspection of all parts using indicators, gauges and checking fixtures.

* Consistently maintained quality and exceeded production quotas.

Additional jobs included:

Spot welder
Fork-lift operator
Landscape crew leader
Delivery driver

MILITARY HISTORY:

U.S. Marine Corps — Honorable Discharge

HOBBIES AND INTERESTS:

Woodworking, swimming and boating

VOLUNTEER WORK:

Lifesaving instructor - Youth Camp
Water Safety instructor

References Upon Request

GERRY M. TURNER

4042 LeClare Street
Milwaukee, Wisconsin 53225
(414) 876-9089

EMPLOYMENT HISTORY:

Power Drive Corporation, 2098 Sandhill Drive, Milwaukee, WI
August 1974 - Present, (available due to downsizing).

ELECTRICIAN
Member of a service-oriented team performing a variety of maintenance functions for a manufacturer of piston rings and component parts.

Responsible for installing and testing new equipment. Perform electrical maintenance on equipment, including those with programmable controllers. Periodically perform preventive maintenance functions.

* Designed and installed small systems for specialized use.

* Qualified blueprint and schematic reader.

* Skilled in the use of electrical testing instruments.

Electrical Worker's Local 275, Oakdale, Wisconsin
August 1967 - August 1974

JOURNEYMAN INSIDE WIREMAN
Performed all types of electrical work: Commercial, Residential and Industrial, including both new installation and maintenance.

* Evaluated as a conscientious and hard-working employee.

* Completed Electrician Apprenticeship.

* Trained aspiring electricians toward Journeyman status.

EDUCATION:

Oakdale Community College
Oakdale, Wisconsin
Completed the 4-year theory portion of Electrician Apprenticeship.

ABC Electric
Oakdale, Wisconsin
Completed course in "Allen Bradley Programmable Controllers."

INTERESTS:

Bowling * Camping * Boy Scout Leader

MILITARY HISTORY:

U.S. Army, Honorable Discharge

References available upon request

WILLIAM EARL WRIGHT

1415 Memphis Drive
Fort Wayne, IN 45837
(312) 387-4903

WORK HISTORY:

Midwest Industrial, Inc. 1523 RR#3, Fort Wayne, IN 45843
1972 - Present (available due to plant closing).

MILLWRIGHT
Responsible for installation, repair and maintenance of all industrial piping and a variety of related equipment including: dust collectors, oil filters, pumps, hoists, trolleys, radiators, and heaters.

Ability to work from blueprints, layouts, directions and specifications as required. Demonstrated skill at problem-solving and planning most efficient use of manpower and machinery.

Supervised up to 35 workers and trained over 50 during career. Directed, planned and performed the installation of new equipment when plant was upgraded.

* Commended by supervisors for skill in handling difficult or dangerous situations.

* Excellent attendance and safety records.

* Quality Circle Leader - 1981

American Manufacturing, Inc., Milwaukee, Wisconsin 53225
1971 - 1972

SECURITY GUARD
Responsible for providing security for 250,000 square foot facility. Check credentials of persons and vehicles entering and leaving the premises. Maintain routine patrol on foot and also monitor various electronic security systems. Maintain radio contact with other guards. Check all windows and doors, see that no unauthorized persons remain after working hours, and insure that fire extinguishers, alarms, sprinkler systems, furnaces, and various electrical and plumbing systems are working properly.

Other skills include: alarm system installation, small appliance repair and general maintenance.

EDUCATION:

Cathedral High School Ft. Wayne, IN	Graduated Industrial Arts

HOBBIES:

Gardening * Horseback riding * Fishing

VOLUNTEER WORK:

Boy Scouts of America - Leader
American Cancer Society - fundraiser
Elks Club - member

MATTHEW HANSON

3543 Mapleton Road
Oakdale, Illinois 58765
(363) 378-3783

EMPLOYMENT HISTORY:

American Lorax, Inc., 3743 Stanley, Oakdale, IL
1979 - Present. Available due to plant closing.

SHIPPING AND RECEIVING COORDINATOR

Supervised 7-10 employees involved in the shipping of parts and the receipt of raw materials for production. Converted manual procedures to computerized system in less than a six-month period. Developed training program for employees working in coordination with Systems Analyst.

MATERIAL HANDLER

Responsible for material handling and warehousing functions required in receiving, processing, storing and shipping of parts, materials, equipment and supplies.

Maintain related clerical records and perform specialized clerical duties as required. Verify quantities, count parts, fill orders, and package parts.

Operate and service mobile equipment, report defects and request needed repair.

MAINTENANCE LABORER

Responsible for general maintenance of buildings, equipment and grounds.

Familiar with various hand tools, power tools, and the operation of mobile equipment. Service mobile equipment and report necessary repairs. Maintain related records.

MILITARY HISTORY:

U.S. Army
Honorable Discharge

EDUCATION:

Oakdale Community College
Oakdale, IL

Introduction to Computers
Basic Business

Taylor High School
Oakdale, IL

HOBBIES AND INTERESTS:

Fishing   Hunting   Gardening   Golf

PERSONAL DATA:

Married
Two children

Excellent health
Willing to relocate

GEORGE M. DAY

1298 Main Street
Henry, Ohio 38298
(932) 837-0987

EMPLOYMENT HISTORY:

Techline, Inc., 427 Maple Street, Henry, Ohio 38298.
1982 - Present (due to corporate reorganization).

TRAFFIC EXPEDITER

Responsible for tracing and expediting urgent material through carriers to ensure timely delivery. Investigate and resolve shortage or damage of delivered material and expedite shipment of rejected material to supplier. Maintain appropriate records of damage.

Related duties include: controlling movement of inbound trucks to avoid demurrage charges, reviewing demurrage charges to ensure accuracy. Maintain and update logs according to ICC regulations. Resolve grief on freight bills from freights payable, prepare Bills of Lading, route outbound shipments and classify material.

* Excellent work and attendance records

* Cited by management for outstanding accuracy and attention to detail

Prior positions included: Fork-lift Operator, Material Handler, Machine Operator, and Shipping Clerk

EDUCATION:

Miami University Oxford, Ohio	Business Administration Industrial Management
Midvale High School Troy, Ohio	College Prep. Curriculum Graduated with honors

CLUBS:

Boy Scouts of America - Leader for six years
Member and Treasurer of Jaycees

HOBBIES AND INTERESTS:

Camping, hiking, fishing and softball

PERSONAL DATA:

Excellent Health * Married * Three Children

Willing to travel and/or relocate.

References Available Upon Request

JOHN P. CLARKE

3197 North Park Avenue
Cumberland, Illinois 57643
(312) 689-9034

PROFESSIONAL EXPERIENCE:

American Distribution, Inc. 38 South Ave., Cumberland, IL
1971 - Present

WAREHOUSE MANAGER
Responsible for supervising and coordinating the activities of over 220 employees for a major distributor of appliance parts.

Developed procedures to coordinate efficient flow of materials throughout the warehouse.

Reorganized various storage areas to accommodate 15% increase in inventory.

Applied knowledge of plant and material-handling equipment to develop training procedures and safety standards for workers in shipping and receiving.

Streamlined verification of paperwork while reducing errors on shipping notices and routing slips.

* Reduced absenteeism over last three years by 9%.

* Initiated employee participation program.

* Excellent rapport with employees.

MATERIAL-HANDLING SUPERVISOR
Supervised workers engaged in the loading and unloading of trucks and freight cars. Reviewed shipping notices, maintained records of incoming parts. Scheduled and directed employees on the safe and efficient operation of equipment.

Other positions included: Shipping clerk, Fork-lift operator

MILITARY:

United States Army - Stateside and Abroad.
Honorable Discharge; attained rank of Corporal.
Duties included: Combat Engineer, Lineman, and Supply Sergeant.
Received various commendations.

TRAINING:

Safety & Supervisor Training, 1976
O.S.H.A. & American In-House Training

HOBBIES AND INTERESTS:

Golf * Fishing * Hunting
Boy Scouts of America - Treasurer 1986, 1987, 1988

PERSONAL DATA:

Married * Two children * Excellent health

HARDY J. CLEMENS

1957 Murray Street
Mt. Clemens, Michigan 48043

EMPLOYMENT EXPERIENCE

Davidoff Door Systems 3991 E. Maple, Birmingham, MI 48009
June 1970 to present

Shipping Supervisor
Supervised and coordinated the efforts of up to 20 individuals in shipping and receiving, warehousing, packaging, and movement of materials for a manufacturer of garage doors. Responsible for personnel issues in the department including promotions, disciplinary actions, and first step grievance procedures. Directed fork lift drivers, warehouse personnel, shippers, truck drivers, material handlers, and a secretary in the performance of their duties.

Maintained all associated records of merchandise shipped and received, including export shipments. Supervised proper loading procedures to effectively eliminate damage in transit. Responsible for all departmental budgeting and purchasing, including the ordering of dunnage materials. Contacted carriers to coordinate shipments and incoming deliveries, communicated arrival times of deliveries to local customers. Interfaced with production control, production line supervisor, customer service department, and vendors to ensure that customers were apprised of schedules and to expedite timely shipments. Calculated truck loads to effectively use available space. Proficient in the use of the IBM 38 computer system to perform all record-keeping functions, as well as calculation of loads.

Other functions included maintaining time cards and attendance records, conducting monthly safety and communication meetings, and monitoring inventory control.

- Instrumental in reducing freight costs by $325,000 and department budget by $400,000 in 1990.
- Increased shipping output per person 7%.
- Implemented a new shipping procedure resulting in a cost savings of $90,000.
- Created a loading diagram which increased the use of cubic space on trucks.
- Requested by the company to train six warehouse personnel at a new manufacturing plant in Texas.
- Commended by management for superb customer relations.

Manufacturing Supervisor
Supervised up to 31 personnel on three production lines in the assembly, repair, and painting of metal-clad entry doors. Scheduled all work assignments, monitored quality, maintained production and downtime reports, and instructed employees. Ensured incoming materials were available for production and worked closely with production control and group leaders to assemble and finish quality products.

- Increased production output 20% on the paint line assembly system.
- Improved output per person in the repair department 40%.
- Started in company as shipper and promoted after four years to manufacturing supervisor, and then to shipping supervisor.

EDUCATION AND TRAINING

Completed various workshops through the company including:

- Interactive Management (DDI)
- Methods Time Management
- M.O.S.T. Standards

Oak Park High School, Oak Park, Michigan: High School Diploma

Epilogue

So you have a new job. Great! Your resume helped you get the interview. You aced the interview. You're ready to get started.

Looking for a new job when you're over 50 isn't much fun. (In fact, it isn't much fun for anyone, no matter what the age.) But you've accomplished what you set out to do, so now you decide you can put all the "job hunting" activity behind you. Now that you're starting a new job, you want to put aside your concerns about discrimination against the over-50 job hunter (real or imagined), the rejection letters, the unreturned phone calls.

But stop and consider. Your years of experience and your maturity tell you, "It ain't over 'til it's over." I recommend three important closing steps:

1. Express your thanks to the people who helped.
2. Reflect on what you've learned from the job-hunting experience.
3. Update your resume and your job-search records.

EXPRESSING YOUR THANKS

You talked to many people during your job search: friends, business contacts, even people you met for the first time. You asked for their assistance, and they provided it. You really don't know how much (or how little) assistance each one provided, so thank them all.

Why?

➤ Simple courtesy. You know how much you valued a thank you during the many years you've been dealing with people.

- Good business. Your experience, especially the experience of needing to find a new job past the age of 50, has taught you that you can never be certain about what the future holds for you. This may not be your last job search. You may need them again. You've learned the importance of "closure"—letting your contacts know what's happened to you.

How?

- A phone call, especially to the people who most directly helped you, followed by a note.
- At the very least, a thank-you card.

UPDATING YOUR RESUME AND YOUR JOB SEARCH RECORDS

Will you need to hunt for a new job again? Of course, you hope not. But in my own case, I changed jobs *twice* after age 50.

The majority of job hunters can't find any record of previous job searches, and they have to start all over again when they're searching for a new job. Just in case this might happen to you someday, file all your logs, cover letters, resumes—even rejection letters—where they will be easily accessible.

RESUME UPDATING

Remember all the time you spent inventorying your skills and accomplishments? How hard you worked on writing your resume? You *may* need to write a new resume some day. My recommendation: Keep a work journal.

Once a month, set aside an hour or so to make a record of every accomplishment, new skill, cost-cutting idea, productivity improvement, sales record, persons trained, courses completed, and so on. Once a year, write a summary of the year's work journals.

Or, write a brand-new resume. You'll never know what challenges and exciting possibilities are awaiting you up the next rung of the career ladder.

I'm a perfect example. I changed jobs twice after age 50: The first time, I went from being a vice-president of personnel for a small supermarket chain to director of personnel for a large drugstore chain. Both jobs were challenging and rewarding—but they cannot be compared with my last job change, when I redirected my career and joined the Transition Team.

I was starting over at a time when most people our age are getting ready to step aside, and I found the experience not exhausting, but exhilarating. Today, I'm president of the company and living proof that the only limits to our achievements are the obstacles we erect ourselves.

Suggested Readings

Allen, Jeffrey G., *Get the Interview,* New York: John Wiley & Sons, 1990.

Allen, Jeffrey G., *The Perfect Cover Letter,* New York: John Wiley & Sons, 1989.

Allen, Jeffrey G., *The Perfect Job Reference,* New York: John Wiley & Sons, 1989.

Allen, Jeffrey G., *Win the Job,* New York: John Wiley & Sons, 1990.

Banning, Kent, and Ardelle Friday, *How to Change Your Career,* Lincolnwood, IL: Career Horizons, 1991.

Berland, Theodore, *Fitness for Life: Exercise for People over 50,* Washington, DC: American Association of Retired Persons.

Berne, Eric, *What Do You Say after You Say Hello,* New York: Bantam Books, 1975.

Bolles, Richard N., *The Three Boxes of Life and How to Get Out of Them,* Berkeley, CA: Ten Speed Press, 1981.

Bolles, Richard N., *What Color Is Your Parachute? A Practical Manual for Job-Hunters and Career-Changers,* Berkeley, CA: Ten Speed Press, 1992.

Boyer, Richard, and David Savageau, *Places Rated Almanac,* Englewood Cliffs, NJ: Prentice-Hall, 1989.

Byrne, John A., *The Headhunters,* New York: Macmillan, 1986.

Dictionary of Occupational Titles, Washington, DC: U.S. Dept. of Labor, 1987.

Franchise Opportunities Handbook, Washington, DC: U.S. Dept. of Commerce, 1991.

Holtz, Herman, *How to Succeed as an Independent Consultant*, New York: John Wiley & Sons, 1983.

How Work Affects Your Social Security Checks, Washington, DC: Social Security Administration, 1992.

Hyatt, Carole, *Shifting Gears*, New York: Simon & Schuster, 1990.

Jukes, Jill, and Ruthan Rosenberg, *I've Been Fired, Too!* Toronto: Stoddart Publishing Co., 1992.

Kennedy, Marilyn Moats, *Salary Strategies*, New York: Rawson Wade Publishers, 1982.

Krannich, Ronald L., and Caryl Rae Krannich, *Salary Success*, Woodridge, VA: Impact Publications, 1990.

Maltz, Maxwell, *Psycho-Cybernetics*, New York: Pocket Books, 1969.

Marsh, Deloss, *Retirement Careers*, Charlotte, VT: Williamson Publishing, 1991.

Molloy, John T., *Dress for Success*, New York: Warner Books, 1975.

Molloy, John T., *The Woman's Dress for Success Book*, New York: Warner Books, 1978.

Occu-facts, Key Largo, FL: Careers, Inc., 1990.

Occupational Outlook Handbook, Washington, DC: U.S. Department of Labor, 1991.

Ray, Samuel N., *Job Hunting After 50*, New York: John Wiley & Sons, 1991.

Saltzman, Amy, *Downshifting*, New York: HarperCollins, 1991.

Schiffman, Stephan, *The Consultant's Handbook*, Boston: Bob Adams, Inc., 1988.

Strasser, Stephen, and John Sena, *Transitions: Successful Strategies from Mid-Career to Retirement*, Hawthorne, NJ: Career Press, 1990.

Yates, Martin John, *Knock 'Em Dead*, Boston: Bob Adams, Inc., 1985.

Zoghlin, Gilbert G., *From Executive to Entrepreneur*, New York: AMACOM, 1991.

Index